Hydrogen Production for a Hydrogen Economy

Topical Conference at the 2009 AIChE Annual Meeting

Nashville, Tennessee, USA
8-13 November 2009

ISBN: 978-1-61567-917-1

Printed from e-media with permission by:

Curran Associates, Inc.
57 Morehouse Lane
Red Hook, NY 12571

Some format issues inherent in the e-media version may also appear in this print version.

Printed by Curran Associates, Inc. (2010)

For permission requests, please contact AIChE
at the address below.

AIChE
3 Park Avenue
New York, NY 10016-5991

Phone: (203) 702-7660
Fax: (203) 775-5177

www.aiche.org

Additional copies of this publication are available from:

Curran Associates, Inc.
57 Morehouse Lane
Red Hook, NY 12571 USA
Phone: 845-758-0400
Fax: 845-758-2634
Email: curran@proceedings.com
Web: www.proceedings.com

TABLE OF CONTENTS

Shaping Our Energy Future: Advanced Research at the U.S. Department of Energy

Jacques Beaudry-Losique, Department of Energy, Washington, DC

The American Recovery and Reinvestment Act of 2009 provided more than $60 billion in clean energy investments. Building upon its record of successful research and development, the Department of Energy is focusing its efforts on meeting the energy objectives of the Administration, including new job creation, breaking our dependence on oil, increasing domestic energy production, promoting energy efficiency, closing the carbon loop-hole, and ensuring U.S. competiveness. This presentation will discuss research and development as well as partnership activities the DOE Office of Energy Efficiency and Renewable Energy is performing to deliver diverse, clean, reliable and affordable energy technologies to the marketplace.

See more of this Session: 2009 Annual Meeting Plenary Session - Looking Forward - Energy Policy & Technology
See more of this Group/Topical: Liaison Functions

The Challenges of Scale and Sustainability in U. S. Energy Policy

Marilyn Brown, Public Policy, Georgia Institute of Technology, Atlanta, GA

The pace of technology improvement and deployment could significantly reduce the cost of stabilizing greenhouse gas concentrations and improve the nation's energy security. Given the need for a large-scale energy transformation, the challenge is to move toward actions that go beyond technology R&D to strategies that target the rapid and large-scale absorption of clean energy technologies into the economy. However, there are numerous barriers to the commercialization and deployment of technology innovations; those that are imposed by administrators, legislatures, and regulators are particularly of interest as they operate at cross-purposes with the U.S. government's GHG reduction goals. This talk focuses on these public policy issues, particularly the challenges of scale and sustainability.

See more of this Session: 2009 Annual Meeting Plenary Session - Looking Forward - Energy Policy & Technology
See more of this Group/Topical: Liaison Functions

The Nation's Renaissance in Energy Technology

Dana C. Christensen, Energy & Engineering Sciences Directorate, Oakridge National Laboratory, Oak Ridge, TN

The nation is facing a renaissance in energy technology to include dramatic changes in transportation, electricity production, distribution and use, and energy efficiency. The nation is also rapidly expanding its scientific capabilities and is currently leading the world in areas such as material science, characterization capabilities, and computational science. Today's industrial challenges are increasingly complex and require the integrated use of these tools in order to solve the critical problems in an economical fashion. In fact, it is becoming increasingly clear that science and technology today requires an integrated approach that includes theory, experimentation and computational science. This new three legged stool for science will be discussed and a number of examples involving the changes in energy technology will be presented.

See more of this Session: 2009 Annual Meeting Plenary Session - Looking Forward - Energy Policy & Technology
See more of this Group/Topical: Liaison Functions

Clean Coal Technology and CO2 Capture: A Perspective

L.-S. Fan, Chemical and Biomolecular Engineering, The Ohio State University, Columbus, OH

Absolute and per-capita energy consumption is bound to increase globally, leading to a projected increase in energy requirements of 50% by 2020. The primary source for providing a majority of the energy will continue to be fossil fuels. However, an array of enabling technologies needs to be proven for the realization of a zero emission power, fuel or chemical plants in the near future. Opportunities to develop new processes, driven by the regulatory requirements for the reduction or elimination of gaseous and particulate pollutant abound.

This presentation will describe the chemistry, reaction mechanisms, particle technology, system engineering, process economics, and regulations that surround the modern utilization of fossil energy, particularly, for the clean coal technology. The presentation will illustrate the salient features pertaining to the fundamental and applied characteristics of the state-of-the-art technologies in practice as well as emerging technologies in development. Emerging technologies evolve largely from the urge for CO2 emission control in fossil energy conversion systems. Further, novel gasification systems based on the chemical looping concepts as well as the calcium looping process for CO2 separation from the combustion flue gas stream will be elucidated in the context of the looping particle design, process heat integration, energy conversion efficiency and economics.

See more of this Session: 2009 Annual Meeting Plenary Session - Looking Forward - Energy Policy & Technology
See more of this Group/Topical: Liaison Functions

Advanced Biofuels - The Path Forward

W. Densmore Hunter, Catchlight Energy, LLC, Federal Way, WA

Oil prices go up and they come down and our ability to forecast them with any real confidence has not been particularly good. A year ago when oil prices were at record highs and global economies were booming, the case for biofuels seemed clear. The path forward is cloudier today. Are we facing a repeat of the boom and bust cycle that decimated alternative energy development in the 1980s? What are the important differences to prevent another collapse? Four factors - climate change, rising global demand for a constrained supply, the end of easy oil, and the political and economic imperatives to reduce oil imports - are expected to sustain biofuels development, despite the hurdles brought about by the global economic crisis. It's a very different situation from the early 1980s. The push for sustainable energy is real and it's here to stay.

How will advanced (lignocellulosic) biofuels develop? It is a much more technically demanding challenge and it will take longer to unfold than grain based ethanol. While chemically similar as polymers of simple sugars, there are important differences between starch and cellulose. Those differences manifest in the way we have traditionally used them. We eat starch. We build houses out of cellulose. Processes involving hydrolysis and fermentation face far greater hurdles with lignocellulosic feedstocks than with starch. In fact it's quite likely that no single technology will dominate the 2nd generation biofuels industry. There are many potential pathways involving biochemical, thermochemical and chemical processes. It's too early to pick the winners. Different situations will favor different solutions. The speaker will explore promising developments that are likely to shape the emerging biofuels industry together with some of the challenges and uncertainties. One conclusion that stands out clearly among all the uncertainty is that the chemical engineering profession will play a dominant role in making sustainable biofuels a reality and thereby contributing in a significant way to the resolution of critical global issues.

See more of this Session: 2009 Annual Meeting Plenary Session - Looking Forward - Energy Policy & Technology
See more of this Group/Topical: Liaison Functions

Wind to Hydrogen Demonstration Project: Results and Lessons Learned

Wednesday, November 11, 2009: 12:30 PM-3:00 PM
Magnolia Boardroom B (Gaylord Opryland Hotel)

Kevin W. Harrison, Electricity, Resources, and Building Systems Integration, National Renewable Energy Laboratory, Golden, CO

Gregory D. Martin, Electricity, Resources, and Building Systems Integration, National Renewable Energy Laboratory, Golden, CO

EXTENDED ABSTRACT

Xcel Energy and the U.S. Department of Energy's (DOE) National Renewable Energy Laboratory (NREL) have collaborated to design, install, and operate the Wind-to-Hydrogen (Wind2H2) project. As the largest provider of wind generated electricity in the United States, Xcel Energy is working proactively with NREL to establish and understand state-of-the-art renewable electrolysis equipment and the operation of a renewable hydrogen production facility. Hosted at NREL's National Wind Technology Center (NWTC), the Wind2H2 system was approved for initial operation in March 2007 and is enjoying success as a demonstration project, producing hydrogen directly from renewable energy sources. This unique research-oriented project uses solar and wind energy to produce and store hydrogen. The stored hydrogen can be used both as a transportation fuel and as an energy storage medium, effectively allowing renewable energy to be stored and converted back to electricity at a later time.

The Wind2H2 project is helping researchers understand the hurdles and potential areas for improvement in emerging renewable electrolysis technologies. By allowing engineers to operate and configure an integrated electrolysis facility, this project has enabled the investigation and analysis of hydrogen production, compression, storage, and electricity generation that is providing valuable data that are being used to improve the designs of future renewable electrolysis systems. This first report on the Wind2H2 project provides important guidance to industry and key stakeholders for development of future renewable electrolysis systems.

The Wind2H2 project is the only renewable hydrogen production facility in the world that can operate multiple electrolyzers in any of the following configurations:

1. Grid connected
2. Directly connected from the output of a photovoltaic array to the electrolyzer stack
3. Real-time electrolyzer stack current control based on a power signal from a wind turbine
4. Closely coupled photovoltaic (PV) and wind energy sources to the electrolyzer stack with custom designed and built power electronics.

NREL and Xcel Energy have undertaken the Wind2H2 project with several key objectives in mind. First and foremost, the Wind2H2 project is being used to demonstrate operation of a renewable electrolysis system, allowing researchers to evaluate actual system performance and costs and to identify areas for cost and efficiency improvements. Additionally, the project provides operational experience with a renewable electrolysis hydrogen production facility, enabling project engineers to investigate operational challenges and to explore system-level integration issues and opportunities for performance and cost improvements resulting from system-level optimization. The project also seeks to investigate how to maximize the use of renewable energy resources in renewable hydrogen production systems by optimizing energy transfer from PV arrays and wind turbines to the stacks of commercial electrolyzers. Finally, the project is designed to explore operational challenges and opportunities related to energy storage systems and their potential for addressing electric system integration issues inherent with high penetrations of variable renewable energy resources.

To help enable greater penetration of renewable energy sources, hydrogen production from renewable electrolysis must be cost-competitive. DOE has a target of reducing the cost of central production of hydrogen from wind-based water electrolysis to $3.10/kg by 2012; by 2017 DOE seeks to reduce this cost to under $2/kg.[1] Electrolyzer manufacturers are improving performance and reducing the capital cost of electrolyzer systems. At the same time, the complete renewable electrolysis system, including the renewable power source, electrolyzer, and interfacing power electronics, must be integrated and optimized to improve system performance and lower costs. The Wind2H2 project presents an excellent research platform to investigate these integration and optimization opportunities.

To achieve the objectives of the Wind2H2 project, NREL engineers have been working to complete a number of project tasks. These tasks include:

- Designing, building, and testing dedicated wind- and PV-to-electrolyzer stack power electronics to more closely integrate the renewable energy resources and electrolyzer stacks
- Modeling and simulating renewable electrolysis system performance to enable improved hydrogen production system designs
- Characterizing renewable energy system impacts on commercial electrolyzer technology and their ability to accommodate the varying energy input from wind and PV sources
- Sequencing multiple electrolyzer systems to improve overall system efficiency, responsiveness, and performance with varying renewable energy sources.

---

[1] U.S. Department of Energy, (October 2007). *Hydrogen, Fuel Cells & Infrastructure Technologies Program: Multi-Year Research, Development and Demonstration Plan*. Office of Energy Efficiency and Renewable Energy.

The challenge of renewable electrolysis is designing and implementing systems that can cost-effectively produce hydrogen from renewable sources using streamlined, robust, and efficient processes. When the wind turbine or solar array is co-located with the electrolysis system, more direct connection between the source and the electrolyzer stack is possible. This close-coupling eliminates the need for long distance transportation of electricity and reduces the number of electrical conversions resulting in a more efficient, cost-effective system.

In this report, specifications of the Wind2H2 equipment (electrolyzers, compressor, hydrogen storage tanks, and the hydrogen fueled generator) are summarized. System operational experience and lessons learned are discussed. Valuable operational experience is shared through running, testing, daily operations, and troubleshooting the Wind2H2 system and equipment errors are being logged to help evaluate the reliability of the system. Project engineers have learned a number of valuable lessons from this system operational experience that will lead to improved design, implementation, and operational plans of renewable electrolysis systems. For example, integrated renewable electrolysis systems require that system components from different manufacturers be configured to function smoothly together. In light of this, programmable logic control systems need to be able to communicate with all major elements of the system (e.g., electrolyzers, compressors, power converters, and load transfer switches).

As another primary goal of the project, NREL engineers have investigated how to maximize renewable energy use and optimize energy transfer within the system. These types of system optimization efforts can significantly reduce the cost of renewable hydrogen. In an analysis of the potential improvements to a wind electrolysis system, project engineers estimated that optimized power electronics would result in a cost improvement of seven percent, reducing the cost of hydrogen produced from wind to $5.83/kg from a baseline of $6.25/kg. For reference, DOE has a target for reducing the cost of central production of hydrogen from wind-based electrolysis to $3.10/kg by 2012. To investigate such optimization opportunities, NREL developed multiple power electronics configurations that convert varying electricity from a solar PV array and wind turbines into the electricity used by the electrolyzer stacks directly. These power converters have the added benefit of executing maximum power point tracking (MPPT) from the wind turbine or PV array to enable higher energy transfer to the electrolyzer stack than a simple, unoptimized, direct connect configuration.

Two electrolyzer technologies are being tested and characterized—namely, polymer electrolyte membrane (PEM) and alkaline electrolyzer systems. Efficiencies of the PEM and alkaline electrolyzers were calculated using experimental data from system operation. Electrolyzer system and stack efficiency is reported on both the higher heating value (HHV) basis of hydrogen (39.4 kWh/kg) and the lower heating value (LHV) basis (33.3 kWh/kg). Table 1 summarizes the results of electrolyzer stack and electrolyzer system efficiency measurements. Stack efficiency peaks at low current while the overall electrolyzer system efficiency is greatest when operating at rated stack current.

Table 1: Electrolyzer system and stack efficiency

Efficiency	PEM Electrolyzer		Alkaline Electrolyzer	
	LHV	HHV	LHV	HHV
Stack Efficiency				
Low Current	80% (5A)	95% (5A)	78% (30A)	92% (30A)
Rated Current	63% (5A)	75% (135A)	59% (220A)	70% (220A)
System Efficiency				
Low Current	0% (15A)	0% (15A)	0% (35A)	0% (35A)
Rated Current	49% (135A)	57% (135A)	35% (220A)	41% (220A)

Voltage-current (V-I) characteristic curves were obtained by sweeping direct current (DC) to the electrolyzer stack. These V-I curves are compared with the theoretical stack voltage (1.48 V per cell) adjusted for the nominal electrolyzer operating gas pressure to determine stack efficiency. System efficiency is determined by summing the electrolyzer stack load plus all ancillary electrical loads divided by the amount of hydrogen produced in kilograms (kg) compared with the HHV of hydrogen.

The PV array connected to the electrolyzer stack with and without an intermediary power converter was tested and analyzed. The PV array was configured to supply different input voltages to a step-down, DC-to-DC (DC/DC) power converter over many days of testing. We found that the use of the power converter increases the energy delivered to the electrolyzer stack by 10% – 20%, depending on the PV array input voltage to the power converter. The efficiency of the power converter decreased as the input voltage from the PV array increased. Still, the highest input voltage from the PV array to the power converter provided the maximum energy capture to the electrolyzer stack. In other words, while suffering from the largest difference between PV voltage and electrolyzer stack voltage (ΔV), the configuration provided the largest amount of energy to the stack over a given day.

A 10 kW wind turbine was connected to one of the PEM electrolyzer stacks through an MPPT-alternating current (AC)-to-DC power converter. This configuration represents a non-grid-tied (i.e., stand-alone) configuration closely coupling a wind turbine to the electrolyzer stack. Testing and analysis shows that the AC-to-DC (AC/DC) converter has the ability to both maintain optimal operation of the turbine while delivering power to the electrolyzer stack without a battery link. It eliminates several power electronics conversions inherent in grid-tied electrolyzer configurations as well as a battery and associated maintenance. This AC/DC power converter is undergoing upgrades that are expected to further increase the energy capture from the wind turbine.

Operation of the Wind2H2 system will continue with a focus on achieving project objectives. Future work includes the continued design and testing of optimization strategies and hardware, along with performance evaluation of these improved systems. This includes maximizing the transfer of wind and solar energy by better integrating system components and developing optimized power electronics packages. Improved sequencing of electrolyzer stacks to maximize hydrogen output will be designed, implemented, and tested and these optimized system configurations will be analyzed to determine the performance of the revised systems. Finally, a virtual connection between the 100 kW wind turbine power signal and 30 kW alkaline stack power controller will vary stack current and enable the characterization of the system performance under highly varying conditions.

One of the major tasks going forward will be to determine how the findings of the Wind2H2 project can improve performance and reduce the cost of renewable electrolysis production systems. The results of system optimization efforts, performance measurements, and evaluation data will be used in economic models to better understand how these system improvements can lower the cost of hydrogen produced via renewable electrolysis.

KEY FINDINGS AND LESSONS LEARNED TO DATE

- **System Integration:** More research and engineering design related to renewable electrolysis system integration will improve energy transfer and overall system efficiency and reduce system complexity and capital costs.
    - Development of optimized power electronics packages in particular is a promising area for system-level improvements.
    - Development of open architecture communication protocols between different component manufacturers would greatly simplify system integration efforts.
- **Codes and Standards:** Development of clear and consistent codes and standards will expedite implementation and reduce the cost of renewable electrolysis projects
- **Efficiency Measurements:** Based on the testing of the Wind2H2 project's small electrolyzer systems, PEM electrolyzers were found to be more efficient than the alkaline electrolyzer; counter to expectations.
- **Energy Transfer Optimization:**
    - Analysis shows a potential 7% reduction in the cost of hydrogen if the wind turbine inverters are replaced with DC/DC converters that feed directly to the electrolyzer stacks.
    - For solar PV systems, NREL determined that the use of an optimized power electronics package (called a maximum power point tracking system – MPPT) captured between 10% and 20% more energy than a direct connection to the electrolyzer stack.

IMPLICATIONS FOR ELECTRIC UTILITIES' COMPONENT SUPPLIERS AND HYDROGEN-BASED ENERGY STORAGE SYSTEMS INTEGRATORS

Notwithstanding the previously identified issue that hydrogen-based energy storage system component capital costs and efficiencies must continue to improve to become competitive with other electric utility production and storage options, the following needs must also be addressed:

- Establish open architecture and communication protocols
- Design wind turbines and electrolyzers with each other in mind, to be part of an energy storage system
- Simplify and standardize codes and standards for electrolysis-based energy storage systems
- Optimize energy transfer within the system and eliminate redundant components
- Standardize system sizes and match component sizes. Until standardized designs and component sizes are developed, system integrators need to determine appropriate component sizes to meet the needs of electrolysis-based energy storage projects on a case-by-case basis.

HIGH TEMPERATURE SOLID-OXIDE ELECTROLYZER 2500 HOUR TEST RESULTS AT THE IDAHO NATIONAL LABORATORY

Dr. Carl Stoots, Idaho National Laboratory, Idaho Falls, ID, USA
Dr. James O'Brien, Idaho National Laboratory, Idaho Falls, ID, USA
Dr. Stephen Herring, Idaho National Laboratory, Idaho Falls, ID, USA
Keith Condie, Idaho National Laboratory, Idaho Falls, ID, USA
Lisa Moore-McAteer, Idaho National Laboratory, Idaho Falls, ID, USA
Joseph J. Hartvigsen, Ceramatec Inc., Salt Lake City, UT, USA
Dennis Larsen, Ceramatec Inc., Salt Lake City, UT, USA

Abstract

The Idaho National Laboratory (INL) has been developing the concept of using solid oxide fuel cells as electrolyzers for large-scale, high-temperature (efficient), hydrogen production. This program is sponsored by the U.S. Department of Energy under the Nuclear Hydrogen Initiative. Utilizing a fuel cell as an electrolyzer introduces some inherent differences in cell operating conditions. In particular, the performance of fuel cells operated as electrolyzers degrades with time faster. This issue of electrolyzer cell and stack performance degradation over time has been identified as a major barrier to technology development. Consequently, the INL has been working together with Ceramatec, Inc. (Salt Lake City, Utah) to improve the long-term performance of high temperature electrolyzers. As part of this research partnership, the INL conducted a 2500 hour test of a Ceramatec designed and produced stack operated in the electrolysis mode. This paper will provide a summary of experimental results to date for this ongoing test.

Introduction

In spite of being the most abundant element in the universe, for the most part molecular hydrogen is not available on earth. The US currently produces approximately 11 million tons of H_2 per year, almost entirely from natural gas. World production of H_2 is currently about 42 million tons per year: 48% from natural gas, 30% from oil, 18% from coal, and 4% from electrolysis [1]. About half is used to produce ammonia, which is then used for fertilizer. The other half is used to convert heavy petroleum sources into lighter fractions suitable for use as fuels.

From the above discussion, 96% of the world's hydrogen is produced from fossil fuels. And the demand for hydrogen is increasing at a rate of more than 10% per year. Long-term supplies of nonrenewable resources such as natural gas and oil for hydrogen production are in question, and their prices are volatile. Furthermore, when fossil fuels are used for hydrogen production, significant quantities of greenhouse gas CO_2 are produced as a byproduct. The majority of this CO_2 is currently exhausted to the environment. Obviously this is not a sustainable means of satisfying H_2 demand for the future.

R&D efforts are underway to develop suitable renewable energy powered and/or nuclear powered water-splitting technologies for carbon-free sustainable hydrogen production. Water-splitting for hydrogen production can be accomplished via high-temperature electrolysis or thermochemical processes, using high-temperature nuclear process heat and electricity. In order to achieve competitive efficiencies, both processes require high-temperature operation. High-temperature electrolytic water-splitting supported by nuclear process heat and electricity has the potential to produce hydrogen with an overall system efficiency near those of the thermochemical processes [2], but without the corrosive conditions of thermochemical processes and without the fossil fuel consumption and greenhouse gas emissions associated with hydrocarbon processes. The Idaho National Laboratory (INL) has been researching the use of solid-oxide fuel cell technology to electrolyze steam for large-scale nuclear-powered hydrogen production. The scope of activities includes computational fluid dynamics modeling, process flow sheet analyses, and experimental testing. Scales of experimental testing extend from ~2 watt button cell tests to 15 kW multi-stack tests.

Utilizing a fuel cell as an electrolyzer introduces some inherent differences in cell operating conditions. For instance, the process heat requirements differ: electrolysis is endothermic while fuel cell operation is exothermic. Futhermore, the direction of oxygen flux is opposite in the electrolyzer mode of operation versus in the fuel cell mode. Finally, cell performance degradation tends to be faster in the electrolytic mode than in the fuel cell mode. This issue of electrolyzer cell and stack performance degradation over time has been identified as a major barrier to technology development. Consequently, the INL has been working together with Ceramatec, Inc. (Salt Lake City, Utah) to improve the long-term performance of high temperature electrolyzers. As part of this research partnership, the INL conducted a 2500 hour test of a Ceramatec designed and produced stack operated in the electrolysis mode.

Experimental Testing Facility

A comprehensive discussion of the INL high temperature solid oxide electrolysis bench scale experiment is presented elsewhere [3]. This same facility is used for button cell testing as well as stack testing. The facility incorporates 5 independent button cell / stack testing stations that can be used for steam electrolysis as well as steam / CO_2 coelectrolysis testing. A photograph of the

Figure 1. INL high-temperature electrolysis laboratory.

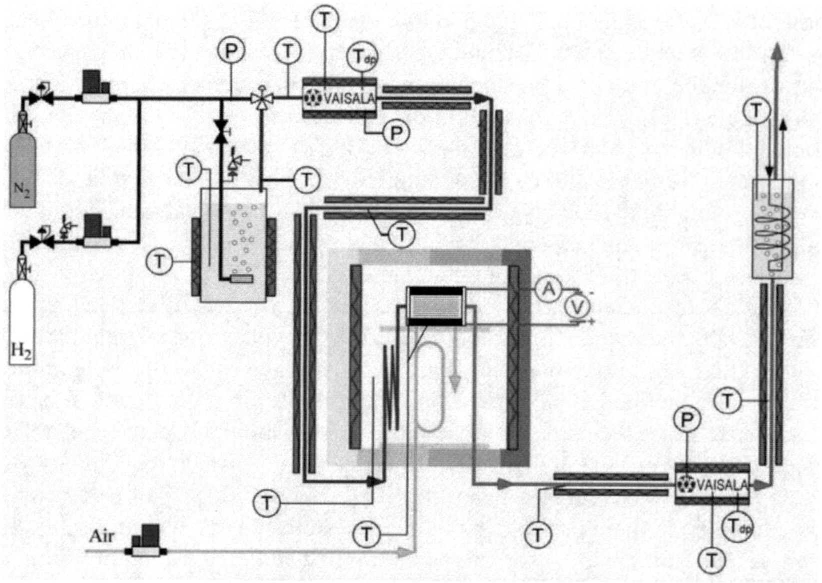

Figure 2. Simplified schematic of INL electrolysis / coelectrolysis stack test apparatus.

laboratory is found in Fig. 1 and a simplified schematic of one stack testing station follows in Fig. 2. Primary components include gas supply cylinders, mass-flow controllers, a humidifier, dewpoint measurement stations, temperature and pressure measurement, high temperature furnace, and a solid oxide electrolysis stack. Nitrogen is used as an inert carrier gas. The use of a carrier gas allows for independent variation of both the partial pressures and the flow rates of the inlet steam, and hydrogen gases while continuing to operate near atmospheric pressure. The flow rates of nitrogen, hydrogen, and air are established by means of precision mass-flow controllers. Air flow to the stack is supplied by the shop air system, after passing through a two-stage extractor / dryer unit.

Downstream of the mass-flow controllers, nitrogen is mixed with hydrogen gas. Hydrogen is included in the inlet flow as a reducing gas in order to prevent oxidation of the Nickel cermet electrode material. The nitrogen / hydrogen gas mixture is mixed with steam by means of a heated water bath. The humidifier water temperature is maintained at a constant setpoint value using computerized feedback control. The dewpoint temperature of the nitrogen / hydrogen / steam gas mixture exiting the humidifier is monitored continuously using a precision dewpoint sensor. Pressure is also measured at the dewpoint measurement stations using absolute pressure transducers. Local stream pressure information is required to determine the mole fraction of steam in the gas mixture at the dew point measurement station. Since the nitrogen and hydrogen flow rates are fixed by the mass flow controllers, and the steam partial pressure is fixed by the bath temperature, the complete inlet gas composition is precisely known at all times. All gas lines located downstream of the humidifier are heat-traced in order to prevent steam condensation. Gas line temperatures are monitored by thermocouples and controlled by means of computer-controlled SCRs.

The electrolysis product stream exiting the furnace is directed towards a second dewpoint sensor upon exiting the furnace and then to a condenser through a heat-traced line. The condenser removes most of the residual steam from the exhaust. The final exhaust stream is vented outside the laboratory through the roof.

Figure 3. 10-cell 10 cm x 10 cm stack mounted on test fixture on furnace base, ready to test.

The rate of steam electrolysis is measured via two different, independent methods: 1) electrical current through the stack, and 2) the measured change in inlet and outlet steam concentrations as measured by the on-line dew point sensors.

The inlet gas mixture is directed to the high temperature furnace, which heats and maintains the electrolyzer at the appropriate test temperature via computer-based feedback control. The furnace also preheats the inlet gas mixture and the air sweep gas. A photograph of a 10 cm x 10 cm 10 cell stack, mounted on its inconel test fixture and resting on the furnace base, is shown in Fig. 3. The stacks were fabricated by Ceramatec, Inc., of Salt Lake City, UT. The stacks have a per-cell active area of 64 cm^2 (10 cm x 10 cm cells) for a total active area of 640 cm^2.

Electrolysis Stack

Planar stacks used for testing by the INL are fabricated by Ceramatec, Inc., of Salt Lake City, UT. The internal components of a typical stack are shown in Fig. 4 and are comprised as follows. The interconnect plate is fabricated primarily from ferritic stainless steel. It includes an impermeable separator plate (~0.46 mm thick) with edge rails and two corrugated flow fields, one on the sweep-gas side and one on the steam / hydrogen side. The height of the flow fields is 1.0 mm. Each flow field includes 32 perforated flow channels across its width to provide uniform gas-flow distribution. The steam / hydrogen flow fields are fabricated from nickel foil. The air-sweep flow fields are made from ferritic stainless steel. The interconnect plates and flow fields also serve as electrical conductors and current distributors. To improve performance, the sweep-side separator plates and flow fields are surface-treated to form a rare-earth stable conductive oxide scale. A perovskite rare-earth coating is also applied to the separator plate oxide scale by either screen printing or plasma spraying. On the steam / hydrogen side of the separator plate, a thin (~10 μm) nickel metal coating is applied.

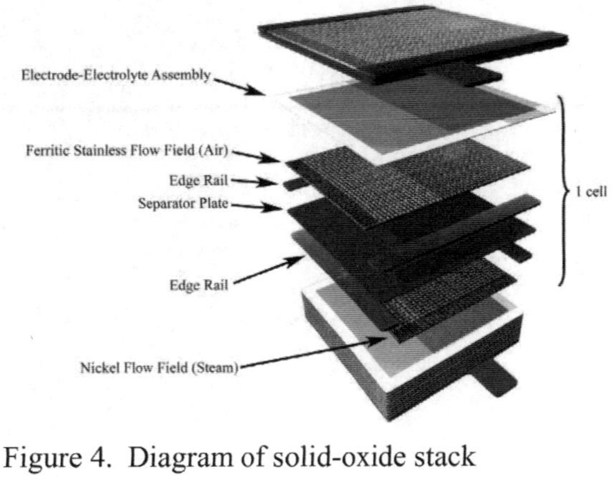

Figure 4. Diagram of solid-oxide stack components.

The electrolyte is partially-scandia-stabilized zirconia, ~140 μm thick. The steam / hydrogen electrode (cathode in the electrolysis mode) is graded, with a nickel cermet layer (~13 μm) immediately adjacent to the electrolyte and a pure nickel outer layer (~40 μm).

The sweep-side or oxygen-side electrode (anode in the electrolysis mode) in previously-tested stacks was composed of a base composition manganite perovskite adjacent to the electrolyte

(14 μm) followed by a graded composition manganite perovskite (22 μm), and a cobaltite bond layer (38 μm). The performance of these earlier cells tended to degrade rapidly in the electrolysis mode.

Post-test examination of the cells used in the INL's three-module 15KW ILS test of 2008 [4] revealed showed that the O_2-electrode delaminated from the electrolyte near the edge. One possible reason for this delamination is excessive pressure buildup due to high O_2 flow in the over-sintered region. Furthermore, electrochemical reactions were also identified as a cause of degradation. Two important degradation mechanisms were examined: (1) transport of Cr-containing species from steel interconnects into the oxygen electrode and cobaltite bond layers in solid-oxide electrolysis cells, and (2) cation segregation and phase separation in the bond layer [5]. To improve cell durability, Ceramatec developed a new spinel coating for the air-side interconnect and flow field to prevent Cr migration. Ceramatec also developed a new cobalt ferrite oxygen electrode and bond layer. These improvements were first tested in-house at Ceramatec. That testing showed that the spinel coating gave a significant reduction in stack degradation, but also resulted in a reduction in initial performance. The new oxygen-side electrode was more stable than the standard oxygen-side electrode. Based upon these promising results, the new coatings and electrode were incorporated into the 10-cell stack test discussed in this paper.

Test Results

Testing commenced with an inlet hydrogen flow rate of 800 sccm, nitrogen flow rate of 1600 sccm, and furnace temperature of 800 C. The inlet dew point temperature was 82 C. This corresponds to an inlet steam mole fraction of 0.6. The theoretical open cell potential for these inlet conditions was calculated to be 0.87 V. The measured open cell potential was 0.86 V.

Fig. 5 presents the results of an initial polarization curve / voltage sweep performed under the above conditions. The voltage was automatically swept from open cell to the thermal neutral voltage of 1.29 volts / cell for the operating temperature of 800 C. The peak current was measured to be 15.1 A, corresponding to a peak current density of 0.24 A/cm^2. Measured per-cell area specific resistance (ASR) values were around 2 Ωcm^2.

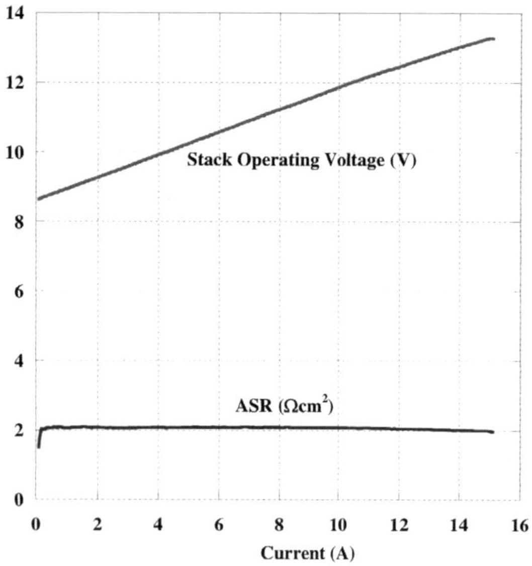

Figure 5. Initial polarization curve for stack.

Long-duration steady-state operations at the thermal neutral voltage of 12.9 V began on May 29 at 16:40. The stack completed 2500 hours at constant input on September 10 at 20:40. A summary graph of the results of the test is shown in Fig. 6. This graph shows the operating voltage, stack current and ASR with data recorded every five minutes. The stack operating conditions were kept constant at the same values as those for the polarization curve test discussed above.

Several events disrupted the steady-state testing. It is not clear whether these events affected the long-term performance of the stack. Most of these disruptions involved interruptions of inlet gas flows (N_2 and H_2), caused by problems with the gas mass flow controller

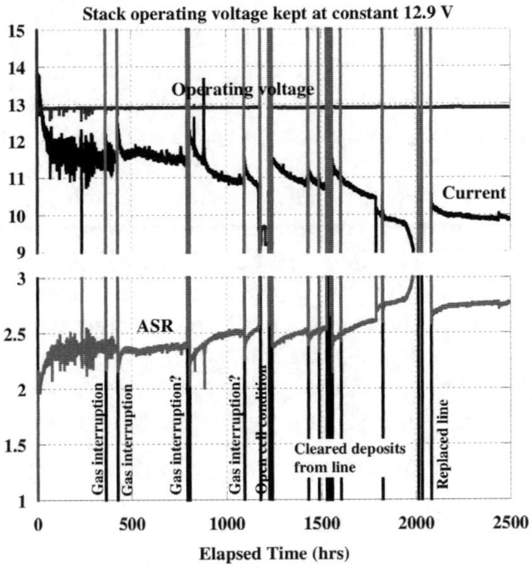

Figure 6. ASR, current, and voltage for the 2500 hour 10-cell stack test.

electronics. Later in testing (beyond 1200 hours), the gas lines in the gas humidifier would become periodically clogged, also interrupting gas flows to the stack. Apparently the laboratory water demineralization system was not operating correctly. A post-test photograph of the hard water buildup within a humidifier line is in Fig. 7. As a result, several analog pressure gauges were installed to monitor the humidifier back pressure such that lines could be replaced before gas flow was interrupted. If blockage of gas flow occurred, the stack would not receive sufficient steam and electrical current would dropped drastically due to insufficient oxygen ions being available to carry the current through the electrolyte.

During the first 60 hours, the nickel oxide in the hydrogen electrode is being reduced and all of the components of the cell are sintering. The ASR of the stack reached 2.3 $\Omega\text{-cm}^2$ at 60 hours elapsed test time, at which point it stabilized. Beyond 60 hours elapsed test time, the average rate of degradation was 8.15 % per 1000 hours.

At the conclusion of 2500 hours the stack had a per-cell ASR of 2.75 Ωcm^2. The hydrogen production rate as measured by the stack current (Faraday's Law) was 690 sccm (41.4 N-liters/hour, 3.7 g/h). The stack electrical current was 9.89 A.

Although the final determination of the cause of this improved performance must await a detailed examination of the cells after they are cooled and the stacked disassembled, it is thought that the spinel coating had the greatest effect on degradation rate. This statement is based upon this test and parametric studies conducted at Ceramatec. The revised formulation of the oxygen electrode had a more minor effect.

Figure 7. Post-test photograph of clogged humidifier line.

Also of interest is that the hydrogen product from the last several 100 hours of testing was intermittently compressed, stored, and subsequently used to produce methane. A methanation experiment was constructed under the INL Hybrid Energy Systems program. This apparatus consists of a ventilated enclosure, hydrogen compressor, hydrogen storage tank, heated methanation reactor, nickel catalyst, micro gas chromatograph, and various other instrumentation. Operating in a semi-continuous fashion, this experiment has successfully demonstrated in-line methanation of electrolysis products, and has provided reaction kinetics information to systems modelers.

Summary

Previous tests of solid-oxide cells as electrolyzers conducted at Ceramatec and at INL using Ceramatec technology have shown high rates of performance degradation (>20% / 1000 hours). Post-test studies conducted at MIT and at Argonne National Laboratory identified oxygen electrode delamination, transport of Cr-containing species from steel interconnects into the oxygen electrode, and cation segregation and phase separation in the bond layer as potential degradation mechanisms. Parametric studies conducted at Ceramatec helped identify improved metallic interconnect coatings and oxygen electrode materials. These materials were incorporated into a 10-cell stack which was subsequently tested for 2500 hours at the INL. After stack stabilization, the stack exhibited a long-duration performance degradation rate of 8.15% per 1000 hours based upon ASR. It is thought that most of the performance improvement resulted from the new metallic interconnect / flow field spinel coating. Overall, the test was quite successful, and demonstrated that good progress has been made in reducing the rate of cell degradation. Finally, the hydrogen product stream from this test was used to demonstrate a new methanation experiment.

Acknowledgements

This work was supported by the U.S. Department of Energy, Office of Nuclear Energy, Nuclear Hydrogen Initiative Program. The Idaho National Laboratory is operated for the U.S. Department of Energy's Office of Nuclear Energy by the Battelle Energy Alliance under contract number DE-AC07-05ID14517.

References

1. Web site http://www.hydrogenassociation.org/general/faqs.
2. Yildiz, B., and Kazimi, M. S., "Nuclear Energy Options for Hydrogen and Hydrogen-Based Liquid Fuels Production," MIT-NES-TR-001, September 2003.
3. Stoots, C., O'Brien, J., Herring, J., and Hartvigsen, J., "Syngas Production via High-Temperature Coelectrolysis of Steam and Carbon Dioxide," *Journal of Fuel Cell Science and Technology*, Vol 6, February, 2009.
4. Stoots, C., Condie, K., O'Brien, J., Herring, J., and Hartvigsen, J., "Test Results From The Idaho National Laboratory 15KW High Temperature Electrolysis Test Facility," Proceedings of the 17th International Conference on Nuclear Engineering (ICONE17), Paper #75417, Brussels, Belgium, July 12-16, 2009.
5. Sohal, M. S., *et. al.*, "Critical Causes of Degradation in Integrated Laboratory Scale Cells During High Temperature Electrolysis," Idaho National Laboratory internal technical report INL/EXT-09-16004, May, 2009.

The RelHy Project: Durability and Efficiency of High Temperature Steam Electrolysis

Florence Lefebvre-Joud, Liten/Dth, CEA-Grenoble, Grenoble, France

Annabelle Brisse, EIFER, Karlsruhe, Germany

Jacob Bowen, Risø DTU, Danish Technical University, Roskilde, Denmark

Bert Rietveld, Energy Research Centre of the Netherlands, Petten, Netherlands

Nigel P. Brandon, Imperial College London, London, UK

High temperature electrolysis has been demonstrated in the 80's by Dornier using tubular ceramic cells as a promising hydrogen production mean. In the recent Hi2H2 project, up-graded SOFC cells, single repeating units (SRUs) and short stacks have been operated in the electrolysis mode leading to results in complete agreement with North American and Japanese ones. Hydrogen production rates up to 0.1 gH2cm-2 hr-1, corresponding to 3.6 Acm-2 at 950°C at the cell level, have been achieved.

However, if these impressing performances confirm the high potential of Solid Oxide Electrolyser Cell (SOEC) technology, they are also associated with high degradation rates compromising market entrance. The RelHy project aims at identifying and overcoming these degradation mechanisms. For such purpose, several instrumented SRUs and short stacks are operated in parallel according to a defined common testing protocol. A special attention is paid to interconnect protective and contacting layers, to sealing and to cell materials. These tests are complemented by extensive post-test characterisations and are simulated by FE-CFD model integrating inputs from micro-modelling in order to take into account microstructure evolution effects in all SRU's materials.

First results obtained with traditional interconnect, protective layer, contact layer, sealing and cell materials are presented and discussed.

See more of this Session: Developments in Electrolytic Routes to Hydrogen
See more of this Group/Topical: Topical 8: Hydrogen Production for a Hydrogen Economy

High Temperature Steam Electrolysis for Hydrogen Production: From Materials Development to Stack Operation

Julie Mougin, Liten/Dth/Lev, CEA-Grenoble, Grenoble, France

Georges Gousseau, Liten/Dth/Lth, CEA-Grenoble, Grenoble, France

Bertrand Morel, Liten/Dth/Lev, CEA-Grenoble, Grenoble, France

François Le Naour, Liten, CEA-Grenoble, Grenoble, France

Florent Chauveau, Université de Bordeaux I, ICMCB-CNRS, Pessac, France

Jean-Claude Grenier, Université de Bordeaux I, ICMCB-CNRS, Pessac, France

Jean-Marc Bassat, Université de Bordeaux I, ICMCB-CNRS, Pessac, France

Fabrice Mauvy, Université de Bordeaux I, ICMCB-CNRS, Pessac, France

High temperature steam electrolysis (HTSE) is one of the most promising options for hydrogen mass production when coupled with nuclear energy. CEA (French Atomic Energy Commission) is carrying out researches in this field, from materials, cells and components developments to stack design and including components and stack testing.

One stack design among those developed at CEA will be addressed in this presentation. This stack is very compact, easy to assemble and simple to operate. Reliability of this design has been demonstrated through several tests of 3 to 5 cells short stacks, cells being commercial 225 cm^2 electrolyte supported cells. Stacks have been operated in pure water vapor (no hydrogen introduced on the cathode side) at 800 to 825°C. Details about performance and durability of this second stacks will be presented.

Besides these activities on stack design, CEA is carrying out researches on new materials and cells, in order to increase the cell performances and the stack performances. For that purpose, alternative materials are studied, and among them layered perovskites for anodes. These researches, carried out jointly by ICMCB-CNRS and CEA highlighted the great potential of nickelates. It has been shown, on button cells at this stage, that the nickelate oxides formulated Nd2NiO4 and La2NiO4 represent a promising alternative to regular LSM anode (*). Such anodes have been deposited on commercial half electrolyte supported cells, and performances up to three times higher have been obtained when operated in HTSE mode compared to an identical commercial cell containing the same cathode and electrolyte but with a conventional LSM anode. Details about the performance of this new alternative anode material will be given in this presentation.

(*) Patent CNRS - EDF (F) "Procédé et dispositif d'électrolyse de l'eau comprenant un matériau oxyde d'électrode particulier", P. Stevens, C. Lalanne, J.M. Bassat, F. Mauvy et J.C. Grenier. France brevet : FR 2872174 2005-12-30

See more of this Session: Developments in Electrolytic Routes to Hydrogen
See more of this Group/Topical: Topical 8: Hydrogen Production for a Hydrogen Economy

Liquid Fuel Production from Biomass via High Temperature Steam Electrolysis

AIChE Annual Conference, Nashville TN, Nov 8-13, 2009
Paper # 405e

Grant L. Hawkes and Michael G. McKellar, Idaho National Laboratory, Idaho Falls, Idaho

Abstract

A process model of syngas production using high temperature electrolysis and biomass gasification is presented. Process heat from the biomass gasifier is used to heat steam for the hydrogen production via the high temperature steam electrolysis process. Hydrogen from electrolysis allows a high utilization of the biomass carbon for syngas production. Oxygen produced form the electrolysis process is used to control the oxidation rate in the oxygen-fed biomass gasifier. Based on the gasifier temperature, 94% to 95% of the carbon in the biomass becomes carbon monoxide in the syngas (carbon monoxide and hydrogen). Assuming the thermal efficiency of the power cycle for electricity generation is 50%, (as expected from GEN IV nuclear reactors), the syngas production efficiency ranges from 70% to 73% as the gasifier temperature decreases from 1900 K to 1500 K. Parametric studies of system pressure, biomass moisture content and low temperature alkaline electrolysis are also presented.

Introduction

The increase of oil prices, the need to reduce the dependence on foreign oil, and the concern for carbon emissions from fossil fuels has created an increased interest in the development of synthetic fuel sources. To reduce carbon dioxide emissions from fossil fuels, biomass to liquid fuel processes such as biodiesel and ethanol have been developed. These processes have an appeal in the sense that the carbon is recycled. However, many issues need to be addressed before the processes become economical. Such issues include biomass collection and preprocessing, maximizing carbon utilization, and process development and optimization.

Only recyclable biomass is considered for this process such as crop residue (corn stover, wheat straw) and wood chips. Typical biomass gasification processes use smaller gasifiers compared to coal gasification due to gathering and transportation costs and issues [1]. Smaller sized units therefore have difficulty justifying the capital cost of air separation units. Therefore these units typically use air as the oxygen source for gasification.

To create a synthetic fuel, the biomass typically needs a hydrogen source. Hydrogen may come by adding steam directly to the gasification process or by steam reforming of natural gas. Direct steam injection has the advantage of adding a source of oxygen to the gasification process. However, additional heat is needed to create the steam. Steam reforming of natural gas also requires heat addition via the combustion of some of the natural gas [2]. The efficiency of producing hydrogen from the methane is about 70% [3].

Many forms of biomass have ash components (inorganic solids) that are very corrosive such as sodium, potassium, and phosphorus which set gasification temperatures below the slagging or melting temperatures of the ash. At these lower temperatures, chemical equilibrium is not achieved and the chemical kinetics of the process becomes important.

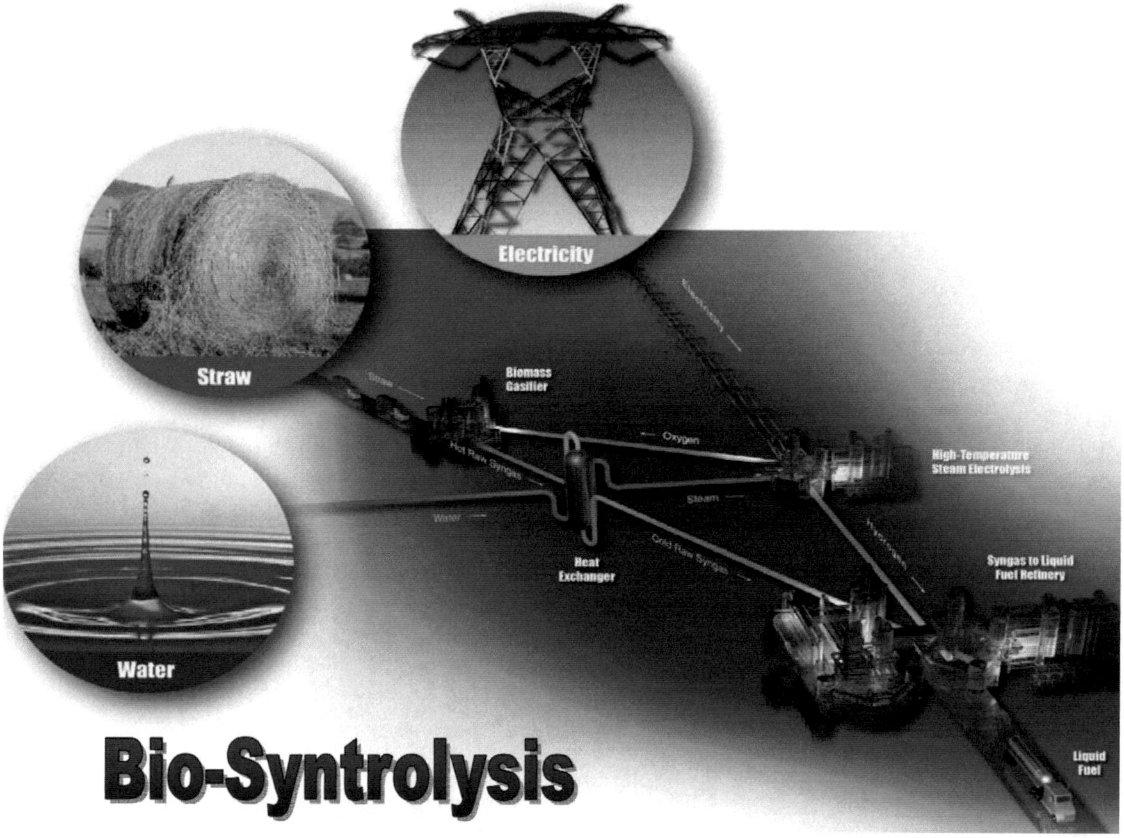

The Idaho National Laboratory is presenting with this paper, a process named Bio-Syntrolysis as shown in the depiction above. This process converts biomass such as barley and wheat straw, corn stover, and wood chips into liquid synthetic fuel to be used in the transportation sector. Unique to this process is the coupling of the biomass gasification heat to the heating of the steam to be used in high temperature steam electrolysis (HTSE). Also integral in this design is the control of the oxidation of the biomass in the gasifier by injecting pure oxygen produced from the electrolyzer. Oxygen is added to maximize the production of Carbon Monoxide (CO) and minimize the production of Carbon Dioxide (CO2). The biomass provides the heat source and the carbon source for synthetic gas production. This process is different from typical biomass to liquid (BTL) gasification processes in that it creates hydrogen from HTSE with an input of electrical power. The source of this electrical power needs to be from a non fossil source such as nuclear, hydro, wind or solar in order to keep this process carbon-neutral. Presented in this paper is a process model flow sheet where syngas is produced. Processing of the syngas to liquid fuels is not modeled in this paper, but is considered in future modeling efforts of the authors. Biomass would be gathered from a radius of about 25 miles as shown in the depiction below. A central non-fossil electrical plant of any type such as light water reactor nuclear, wind, solar, hydro, or Gen IV nuclear would provide the electricity. It would be centered where the biomass is grown to cut down on construction and maintenance costs of power lines. Bio-Syntrolysis plants would convert the biomass into liquid fuel. Inputs would be water, agricultural residue (recyclable biomass), and electricity.

Bio-Syntrolysis: Regional Concept

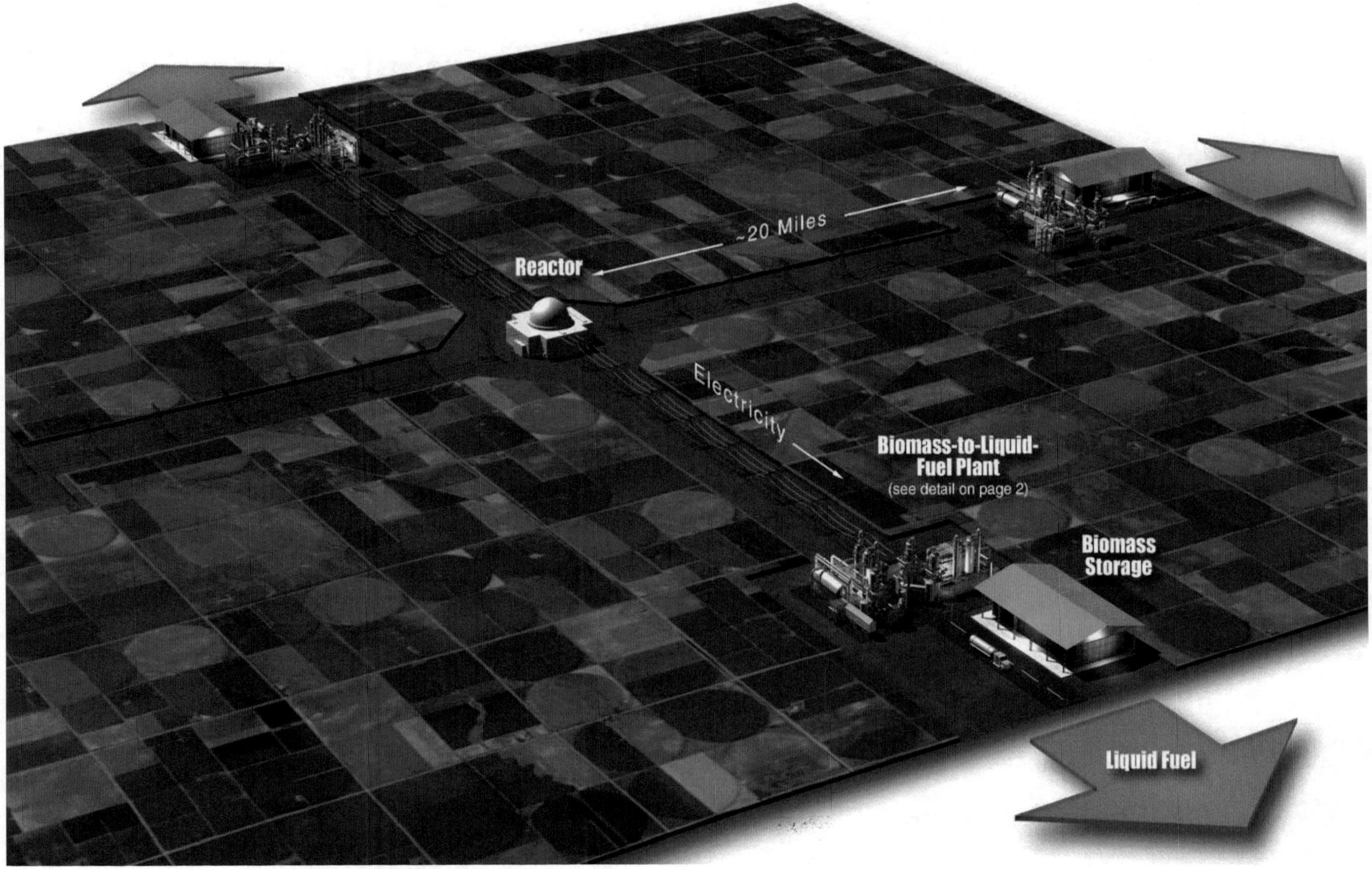

Nomenclature

η_{syngas} Syngas production efficiency
η_{th} Thermal efficiency of power cycle
LHV Molar lower heating value
$\dot{n}$ Molar flow rate
Pwr Power
elec Electricity

Process Description

The process model was developed using the Honeywell UniSim process modeling software. This commercial software is used in the oil and gas industry and can model thermo-chemical systems ensuring chemical, energy and mass balances [4].

Four types of recyclable biomass were considered parametrically for this study. They are barley straw, wood bark, corn stover and switchgrass. The ultimate analysis of the dry, ash free, biomasses is shown in Table 1 and is based on the PHYLLIS database of the Energy Research Centre of the Netherlands [5]. Inputs from the biomass include the lower heating value. For example, the lower heating value of the barely straw is 18,460 kJ/kg. The dry ash is primarily silicon oxide with some potassium and calcium oxide. Both the ash and the dry, ash free, biomasses were modeled as hypothetical components. The composition and heating value of the biomasses are the only necessary components for the gasification process model. The gasifier uses a Gibbs reactor which minimizes the Gibbs free energy to determine the most likely products based on the composition, flow rates, pressure and temperatures of the barley straw, water, and oxygen. A large number of components were considered as products, but only those presented in Table 2 were present.

Table 1. Ultimate analysis of various recyclable biomasses.

Biomass	C (%)	H (%)	O (%)	N (%)	S (%)	Cl (%)	Ash (%)	Water (%)	LHV (kJ/kg)
Barley Straw	49.2	6.0	44.1	0.4	0.06	0.43	4.9	11.5	18460
Bark	52.1	5.8	41.4	0.57	0.03	0.016	3.3	11.5	16029
Corn Stover	49.3	6.0	43.6	0.7	0.11	0.28	5.1	11.5	17746
Switchgrass	53.2	6.4	39.0	1.3	0.11	0.0	10.1	11.5	18653

Table 2. Molar composition exiting gasifier with barley straw as biomass.

Component	% Mole Fraction
Carbon Monoxide	51.81%
Hydrogen	43.86%
Water	2.22%
Carbon Dioxide	0.99%
Nitrogen	0.20%
Hydrogen Chloride	0.16%
Methane	0.73%
Hydrogen Sulfide	242 ppm
Ammonia	24 ppm
Carbonoxysulfide	1 ppm

The primary purpose of modeling the ash is to simulate the heat captured by the ash during gasification, but the ash does not react chemically in the gasifier. A temperature dependent specific heat is determined based on the composition of the ash.

Figure 1 is the process flow diagram for biomass syngas production using high temperature electrolysis (HTE). The process is scaled to a biomass feed rate of 1 kg/s. The biomass is reacted with oxygen from the steam electrolyzer at a variety of gasifier temperatures ranging from 1500 K to 1900 K. The gasification temperatures were selected to ensure chemical equilibrium and are based on temperatures for a Shell gasifier for coal [6]. Slagging of the ash will occur at those temperatures. Just enough oxygen is added to make the gasification process adiabatic.

The gasifier product stream and the hydrogen stream from the electrolyzer exit are cooled in the Heat Recuperator 1 heat exchanger to generate steam at 1073 K for the electrolysis unit. The water entering Heat Recuperator 1 is supplied by a make-up water source that is pumped to pressure and the water recycled from the electrolyzer outlet.

Oxygen from the electrolyzer is sent to the gasifier in a controlled manner. Extra oxygen not needed goes through a cool down process. Enough oxygen is added to the gasifier to keep the process adiabatic. The gasifier product stream is quenched with water to the point of saturation for the purpose of modeling the energy loss due to particulate removal using a quenching process. The water contains some undesirable components from the gasifier including hydrogen sulfide and hydrogen chloride and therefore can not be recycled. The contaminants in this stream are removed in a contaminant removal process that is based on the Rectisol process. Methanol is refrigerated to 233 K and enters the top of an absorber. The gasifier product stream is cooled to remove excess water and enters the bottom of the absorber. The refrigerated methanol captures the hydrogen sulfide, some of the carbon dioxide, the ammonia, and the hydrogen chloride. The methanol is regenerated in a distillation column which is then recycled back to the absorber. Any losses of methanol are replaced. The contaminants exit the distillation column at the top of the column as a sour gas. The heat for the reboiler of the distillation column is provided by combusting a small stream of biomass at a feed rate of 0.025 kg/s (see Figure 2). The energy used to pump and to refrigerate the methanol and the biomass burned to supply heat for the reboiler are taken into account in efficiency and carbon utilization analyses.

Hydrogen from the electrolysis process is mixed with the gasifier stream to produce a syngas of hydrogen to carbon monoxide ratio of 2. This ratio is ideal for many synthetic fuel processes such as the Fischer Tropsch, FT.

The inlet mole composition of the stream at the electrolyzer is 10% hydrogen and 90% steam.

The heat from the gasifier and electricity provide the energy to split the steam into oxygen and hydrogen. The process out stream is composed of 90% hydrogen and 10% steam. The stream is cooled to near ambient temperatures by the Heat Recuperator 1. The water is separated from the hydrogen at the H2/Water Separator. The water is recycled and mixed with the makeup water.

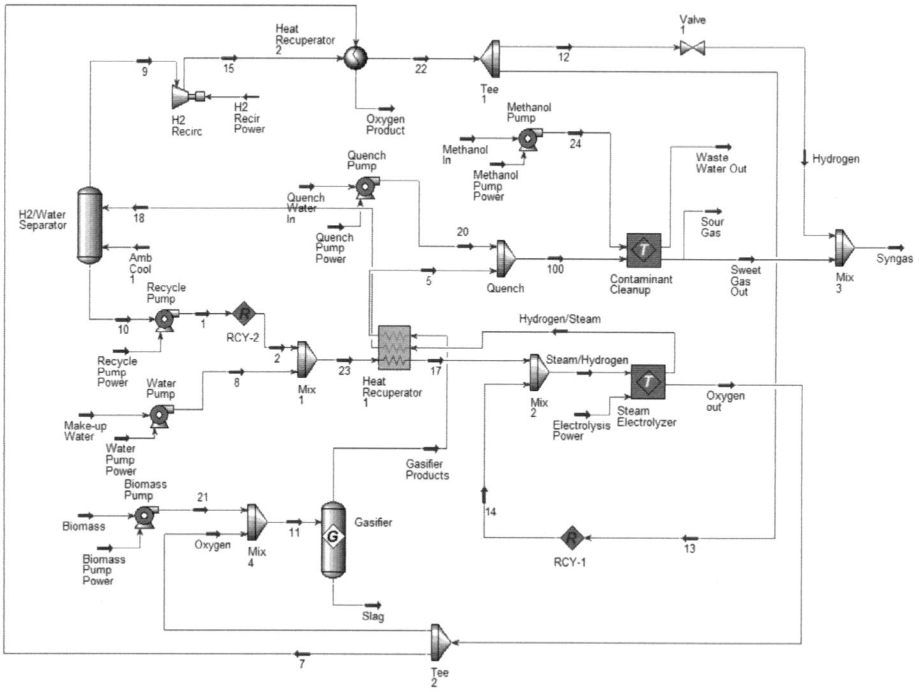

Figure 1. A process flow diagram of biomass gasification with high temperature electrolysis.

The hydrogen exiting the separator is used primarily for syngas production but some is recycled to the inlet of the electrolysis process to provide reducing conditions at the water side electrode.

Pressure losses in the heat exchangers and the gasifier are set at 20 kPa per pass. An average pressure of 2.78 MPa was assumed for the analysis. The isentropic efficiencies of the pumps and hydrogen circulating compressor are set at 75%. The refrigeration compressor has an efficiency of 90%.

The steam electrolyzer is a module developed at the Idaho National Laboratory to simulate the temperature and compositional averaged Nernst potential of an electrolysis process. Details of the module are found in references [7, 8]. Given the number of cells, the area specific resistance (ASR), the current density, the cell area, and the inlet flow conditions of the process stream, the module will calculate the average Nernst potential, the per-cell operating voltage, the electrolysis power, the process heat needed and the outlet flow conditions of the hydrogen and oxygen streams. For this parametric study, it was assumed that the ASR was 0.4 ohms*cm^2, the current density was adjusted to provide thermal neutral conditions (adiabatic and isothermal), and the per-cell area was 225 cm^2. The thermal neutral conditions set the per-cell operating voltage at 1.288 Volts/cell. The number of cells was adjusted until the hydrogen to carbon monoxide syngas ratio was 2.0. Within the module, a setting was made so that 88.9% of the steam entering the electrolyzer was converted to oxygen and hydrogen resulting in an outlet stream molar composition of 90% hydrogen and 10% steam.

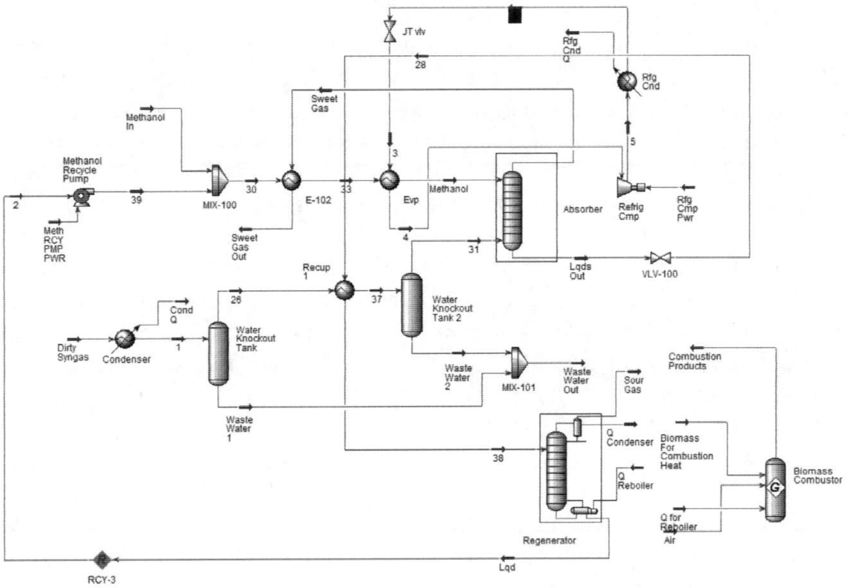

Figure 2. A process flow diagram of the contaminant cleanup process.

Syngas Production Efficiency and Carbon Utilization

Two measurements of process performance were used in this analysis: the syngas production efficiency and carbon utilization. The syngas production efficiency is defined as the thermal value of the syngas divided by the sum of thermal value of the biomass and the thermal equivalent value of the electricity used in the process. The thermal values of the syngas and the biomass are calculated by multiplying the lower heating values of each with their respective molar flow rates. The thermal value of the electricity is found by summing the electric power of the system and dividing it by the thermal efficiency of the electric power cycle providing the electricity. The heat for the gasification process is not included in this efficiency because the process is adiabatic. The molar flow rate of the biomass is the biomass entering the gasifier and the biomass that is combusted to provide heat for reboiler in the methanol regeneration column. The refrigerator compressor power and the methanol recycle pump power are also included in this efficiency.

$$\eta_{syngas} = \frac{\dot{n}_{syngas}\, LHV_{syngas}}{\left(\dot{n}_{biomass}\, LHV_{biomass} + \sum Pwr_{elec} \big/ \eta_{th}\right)} \tag{1}$$

The carbon utilization is the mass flow of carbon in the syngas divided by the mass flow of carbon in the biomass. Only the carbon in the carbon monoxide component of the syngas is considered because all components except for hydrogen and carbon monoxide are considered contaminants or inerts.

$$U_c = \frac{Mass_{C,Syngas}}{Mass_{C,Biomass}} \tag{2}$$

Results

Results are shown in Figures 3 through 9. A baseline case using barley straw is outlined to show the process progression. For this case, the gasifier temperature was set at 1500 K. The composition, pressures, temperatures, and flows of the streams into and out of the process are shown in Table 3. The composition is in % mole fraction unless otherwise indicated. The contaminants and inerts in the syngas are less than 2% of the total mole faction. However the hydrogen chloride and hydrogen sulfide contents are high enough that they need to be removed. The mixing temperature of the hydrogen and the sweet gas out streams is 426 K which is low enough that the water gas shift reaction need not be considered. The syngas is delivered at a pressure of 2.71 MPa and 426 K at a hydrogen to carbon dioxide ratio of 2.0. A Fischer Tropsch syngas to synfuel process with a cobalt catalyst needs a syngas feed stream with a ratio of 2.12, pressure at 2.70 MPa and a temperature of 505 K [9, 10]. Therefore the syngas will need to be heated before entering a FT process.

For the baseline case, the number of cells in the electrolyzer is 53,000. The current density is 0.6193 amperes per cm^2, resulting in a current of 139 amperes. The electrolysis power is 9,500 kW.

The temperature vs. heat flow profiles of the recuperative heat exchangers are shown in Figures 3 and 4. Figure 3 is the Heat Recuperator 1 heat exchanger. The minimum approach is 50 °C and has a log mean temperature difference, LMTD, of 201 °C. One can see the steam generation in the cold profile. Figure 4 shows the profile for the Heat Recuperator 2 heat exchanger. This heat exchanger also had a minimum approach of 50 °C. The LMTD is 195 °C for this heat exchanger.

Table 3. Composition and flow conditions of the inlet and outlet streams.

Molar Composition							
	Biomass (Mass Fraction)	**Make-up Water**	**Syngas**	**Sour Gas**	**Waste Water**	**Sweet Gas Out**	**Hydrogen**
Hydrogen			65.96%	6.44%	28 ppm	45.12%	99.90%
Carbon Monoxide			32.98%	17.56%	36 ppm	53.23%	
Water	11.5%	100%	0.04%		99.97%		0.10%
Carbon Dioxide			0.33%	66.07%	88 ppm	0.56%	
Oxygen				100%			
Nitrogen			0.13%	0.24%	1 ppm	0.20%	
Hydrogen Chloride			744 ppm	6.16%	12 ppm	0.12%	
Hydrogen Sulfide			114 ppm	0.93 %	7 ppm	184 ppm	
Methane			0.46%	2.59%		0.74%	
Ammonia			4 ppm	31 ppm	167 ppm	7 ppm	
Carbonoxysulfide				6 ppm		1 ppm	
Barley Straw	83.6%						
Ash	4.9%						
Flow Conditions							
Pressure (MPa)	0.1013	0.1013	2.71	0.131	2.7	2.71	2.71
Temperature (K)	289	289	426	185	323	319	606
Mass Flow (kg/s)	1	0.690	1.095	.0166	0.117	1.017	0.078

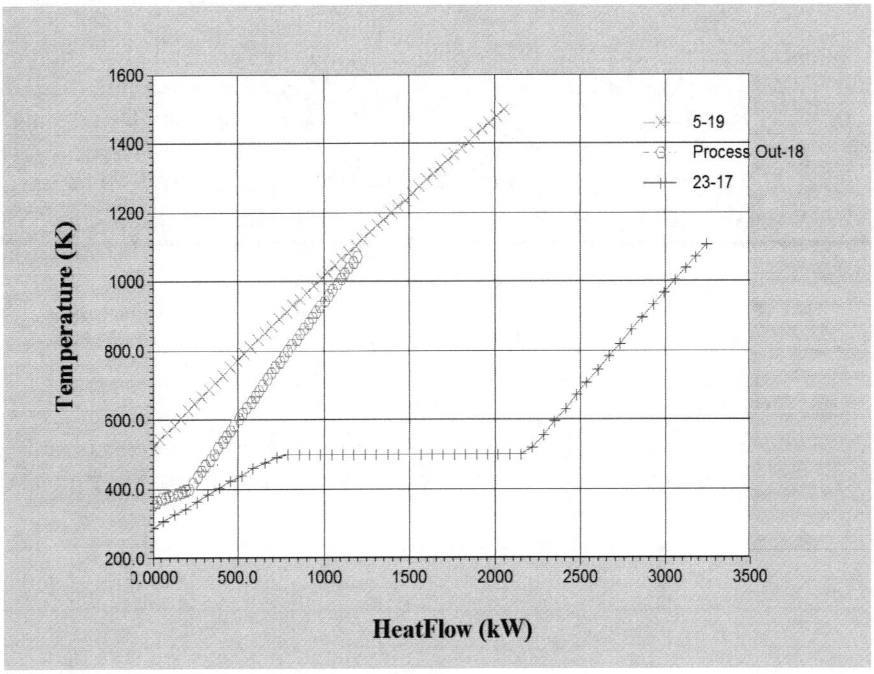

Figure 3. Temperature profile plot for the Heat Recuperator 1 heat exchanger.

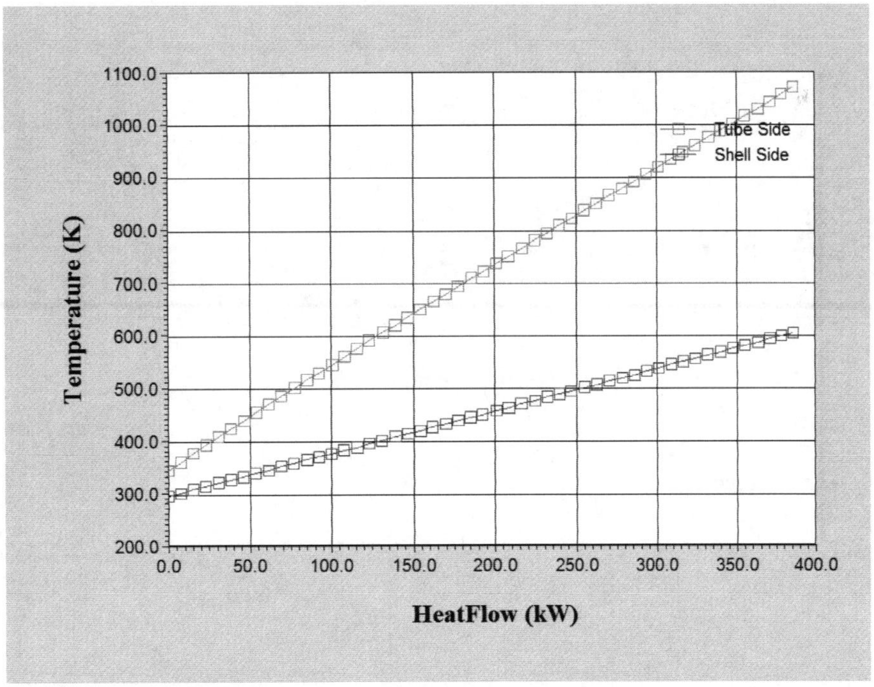

Figure 4. Temperature profile plot for the Heat Recuperator 2 heat exchanger.

The carbon utilization was calculated to be 94.5% for this analysis. In other words 94.5% of the carbon in the biomass was converted to syngas. As stated before this carbon utilized is the carbon found only in the carbon monoxide. Figure 5 shows the carbon utilization and syngas production efficiency as a function of gasifier temperature for the four biomasses considered. Carbon utilization is plotted on the left axis while syngas production efficiency is plotted on the right axis. The carbon utilization goes down with gasifier temperature because more oxygen is needed to make more CO2 and hence more heat to heat the gasifier products. Syngas production efficiency hits a peak at 1600 K due to methane formation, and then decreases with temperature. The same reason exists for the decrease in syngas production efficiency because more CO2 is made and is not available to be made into syngas with constituents of CO and H2.

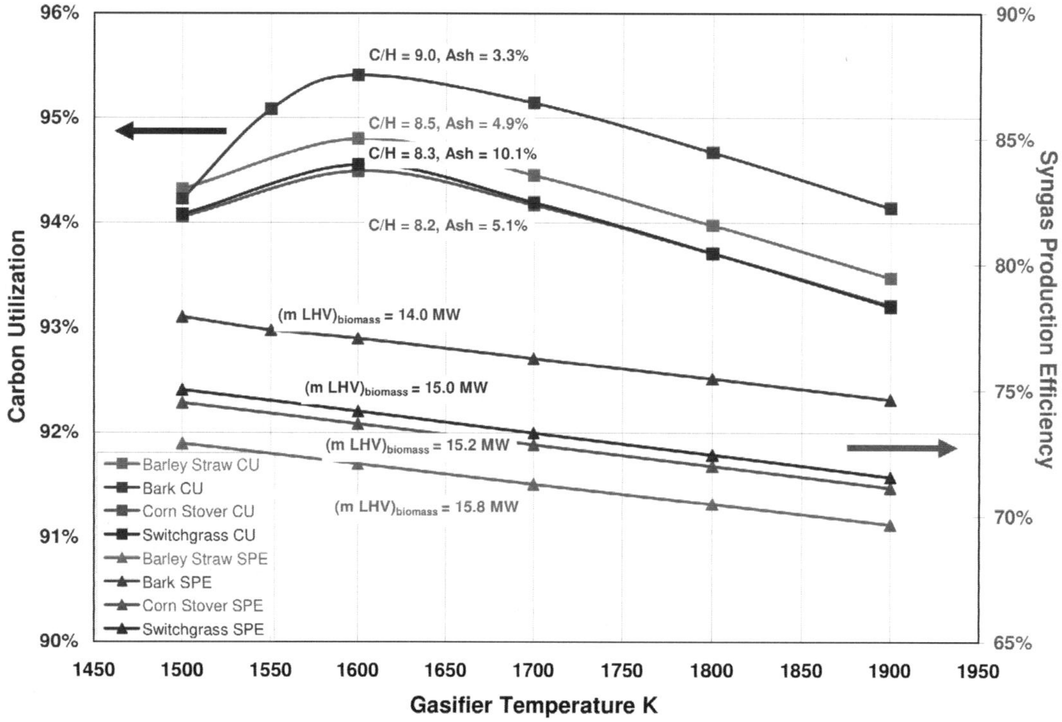

Figure 5. Carbon utilization and syngas production efficiency for various biomasses as a function of gasifier temperature.

Figure 6 shows the carbon utilization and syngas production efficiency as a function of system pressure. Carbon utilization is not affected by lowering the system pressure to 0.345 MPa, but the syngas production efficiency goes down from an average of 73% to 67%. This is due to the electrical cost of compressing the syngas up to 2.73 MPa.

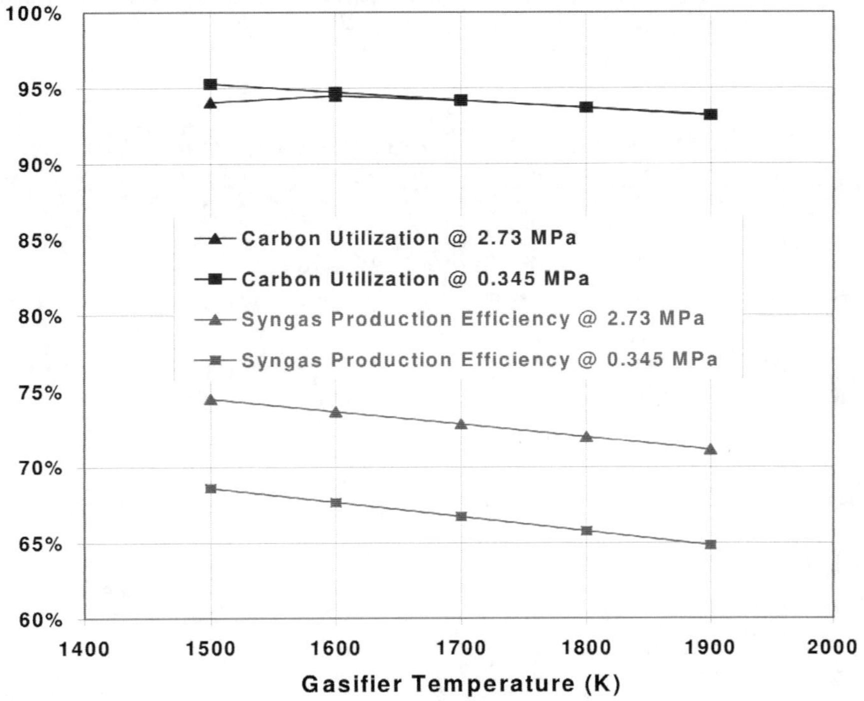

Figure 6. Carbon utilization and syngas production efficiency as a function of system pressure for corn stover.

Shown in Figure 7 is the carbon utilization and syngas production efficiency as a function of moisture content in corn stover at a gasifier temperature of 1600 K. Carbon utilization takes a steep drop as the moisture content increases. This is due to the fact that more oxygen is needed to make more CO_2 to make heat to heat up the extra water in the biomass. The CO_2 is not converted into CO in this instance and the amount of carbon entering the syngas goes down. This process does have the capability of making more hydrogen in this instance to react with the CO_2 and convert it into CO. It just was not implemented or explored for this paper. The syngas production efficiency also decreases as the moisture content increases for the same reason stated above.

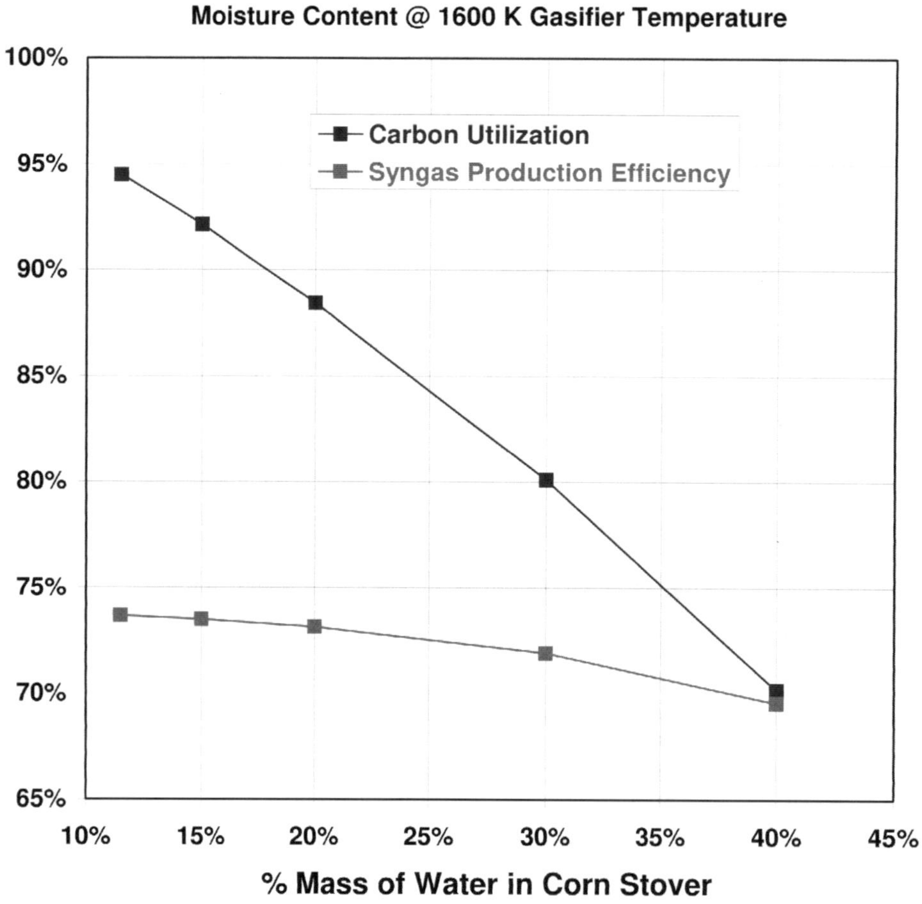

Figure 7. Carbon utilization and syngas production efficiency varying with moisture content in corn stover at a gasifier temperature of 1600 K.

Carbon utilization and syngas production efficiency are plotted in Figure 8 for low temperature alkaline electrolysis (LTE) as a function of gasifier temperature. The carbon utilization is identical for HTSE compared to LTE. The syngas production efficiency for HTSE averages about 73% compared to 57% for the LTE. This is mainly due to the increased electrical requirement for LTE which is about 56% higher. Model results show that it takes 123,000 kW per kg/s of electrical power to electrolyze steam with the HTSE method. The LTE method takes 192,00 kW per kg/s of electrical power for electrolysis of liquid water.

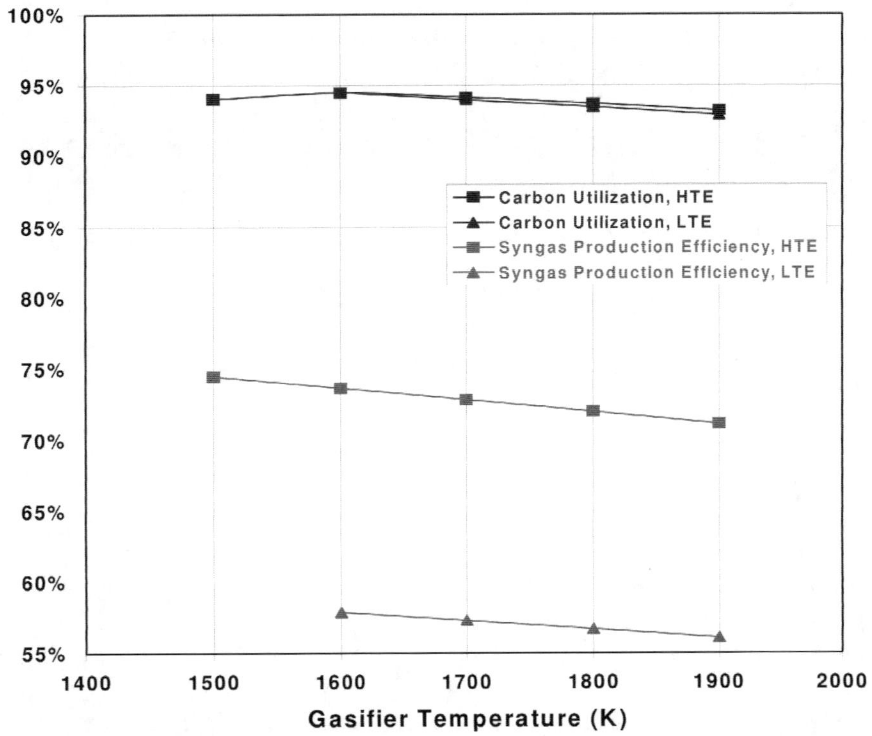

Figure 8. Carbon utilization and syngas production efficiency comparing HTSE and LTE as a function of gasifier temperature.

Figure 9 shows a plot of syngas production efficiency as a function of power cycle efficiency. This efficiency obviously has a strong correlation with the power cycle efficiency since it is in the denominator of the Eq. 1. The purpose of including the power cycle efficiency in the syngas production efficiency is to be able to compare overall system performance where the heat of the power cycle is the ultimate driving force. A power cycle range from 25% to 55% was used to include most of the electric power cycles developed. A very high temperature gas GEN IV nuclear reactor has a thermal power cycle efficiency near 50%. If the reactor were used to generate the electricity for the HTE process, the overall syngas efficiency would be near 73%.

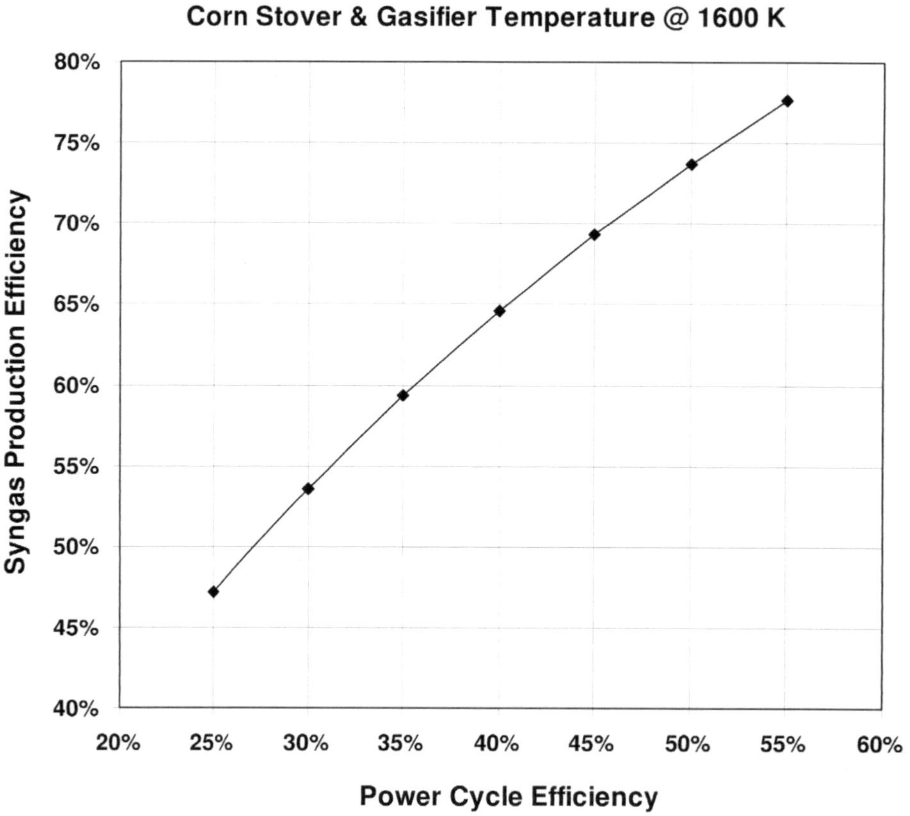

Figure 9. Syngas production efficiency as a function of power cycle efficiency for corn stover with a gasifier temperature of 1600 K.

Economic Analysis

A thorough and comprehensive economic analysis was performed for the Bio-Syntrolysis process using the DOE H2A methodology [11]. This economic analysis was performed with nth of a kind assumptions. The biomass to liquid plant was sized for 23,800 barrels per day. The analysis shows synthetic diesel can be produced for $136/barrel with optimistic assumptions at $116/barrel, and pessimistic assumptions at $225/barrel. Capital cost is roughly $ 2 billion with an operating cost of $ 1 billion per year. The capital cost does not include the electrical production facility, but the operating cost assumes purchasing electricity at $0.05 per kW-hr. The electrical need for a plant this size is approximately 950 MW electric. Biomass would be gathered from a radius of about 25 miles to the Bio-Syntrolysis plant. Production costs for the fuel are at $2.80 per gallon. When incorporating a 13% rate of return, the cost goes up to $3.25 per gallon ($136/barrel).

Conclusions

A process model was developed to simulate the generation of syngas by gasification of biomass and high temperature electrolysis of steam for the process known as Bio-Syntrolysis. Biomass provides the heat source for the HTSE and the carbon source for the syngas. The carbon utilization is affected only slightly by gasifier temperature as the utilization varies between 94% and 95%. The syngas production efficiency is closely tied to the power cycle efficiency. At a power cycle efficiency of 50%, the syngas production efficiency varies from 70% to 73%.

Parametric studies were shown for carbon utilization and syngas production efficiency varying with gasifier temperature, system operating pressure, HTSE versus LTE, biomass moisture content, and four biomasses (barley straw, corn stover, bark, and switchgrass).

An economic analysis was performed for this process showing liquid fuel production at $3.25 per gallon. Although the analysis looks promising, other biomass inputs would change the configuration. If large concentrations of sodium, potassium or phosphorus compounds are found in the ash, the gasifier temperatures would need to be much lower to prevent slagging.

Acknowledgements

This work was supported by the Idaho National Laboratory, Laboratory Directed Research and Development program and by the U.S. Department of Energy, Office of Nuclear Energy, Nuclear Hydrogen Initiative Program.

Copyright Statement

References

1. Higman, Christopher, and Maarten van der Burgt, 2003, *Gasification*, San Diego; Elsevier, pp. 72-75.
2. 2001, Kirk-Othmer Encyclopedia of Chemical Technology, Topic: Hydrogen, New York: John Wiley & Sons, Inc., Vol. 13, pp. 776-777
3. Probstein, R. F., R. E. Hicks, *Synthetic Fuels*, Dover Publications, Inc. Mineola, New York, 2006.
4. UniSim Design, R360 Build 5, Honeywell International Inc., Copyright 2005-2006.
5. PHYLLIS, ECN-Biomass, Version 4.13, Energy Research Centre of the Netherlands (ECN), www.ecn.nl/phyllis.
6. Suidevel, P. L., "Shell Coal Gasification Process for Power and Hydrogen/Chemicals." GTC Conference, Washington D.C., October 2004.
7. O'Brien, J. E., McKellar, M. G., Hawkes, G. L., and Stoots, C. M., "Development of a One-Dimensional Co-Electrolysis Model for Use in Large-Scale Process Modeling Analysis," Fifth International Conference on Fuel Cell Science, Engineering & Technology, June 18-20, 2007, New York, USA.
8. O'Brien, J. E., McKellar, M. G., Stoots, C. M., Herring, J. S., and Hawkes, G. L., "Parametric Study of Large-Scale Production of Syngas via High Temperature Electrolysis," in review, *International Journal of Hydrogen Energy*, 2008.
9. Dry, Mark E., "High Quality Diesel via the Fischer-Tropsch Process -- A Review," 2002, *Journal of Chemical Technology and Biotechnology*, Vol. 77, No. 1, pp. 43-50.
10. Steynberg, A. P., and Dry, M. E., 2004, *Studies in Surface Science and Catalysis, 152, Fischer-Tropsch Technology*, San Diego: Elsevier, pp. 196-257
11. www.hydrogen.energy.gov/h2a_analysis.html

Comparison Between Hydrogen Production Alternatives From Nuclear Energy

Maximilian B. Gorensek, Computational Science and Statistics Department, Savannah River National Laboratory, Aiken, SC

William A. Summers, Hydrogen Technology Center, Savannah River National Laboratory, Aiken, SC

This paper will compare and contrast the alternative methods that have been proposed for the production of hydrogen from nuclear energy. Using performance projections that are available in the literature, and taking alkaline electrolysis with nuclear power as the benchmark, it will compare heat and power requirements on a unit hydrogen production basis. Estimates will also be made for the cost of hydrogen production by each of the alternative routes.

See more of this Session: Nuclear Hydrogen Plant Design, Analysis and Commercialization
See more of this Group/Topical: Topical 8: Hydrogen Production for a Hydrogen Economy

A Systematic Approach towards the Synthesis and Evaluation of Hybrid Alternative Thermochemical Cycles

Ryan J. Andress, Chemical and Biological Engineering, Rensselaer Polytechnic Institute, Troy, NY

Lealon L. Martin, Chemical and Biological Engineering, Rensselaer Polytechnic Institute, Troy, NY

Here we expand on our previous work, a process systems engineering-based methodology for the initial evaluation of alternative thermochemical cycles for hydrogen production. Modifications in the identification algorithm are made to include thermochemical cycles with electrochemical steps, referred to as hybrid cycles, in the reaction cluster search space. The new formulation is demonstrated with a case study on the hybrid copper-chlorine thermochemical cycle. Results from our systematic evaluation suggest the proposed methodology is robust over a wide range of temperatures and candidate species. In this case study, we provide hybrid cycles with estimated base efficiencies of up to 36%.

See more of this Session: Nuclear Hydrogen Plant Design, Analysis and Commercialization
See more of this Group/Topical: Topical 8: Hydrogen Production for a Hydrogen Economy

Heat Pump Cycles to Generate Electricity From Low Temperature Waste Heat

Ramalingam Subramaniam, Chemical & Biomolecular Engineering, Tulane University, New Orleans, LA

John C. Prindle, Chemical & Biomolecular Engineering, Tulane University, New Orleans, LA

Alexandra I. Lupulescu, Chemical & Biomolecular Engineering, Tulane University, New Orleans, LA

Victor J. Law, Chemical & Biomolecular Engineering, Tulane University, New Orleans, LA

This paper presents an extensive study of heat pump cycles and associated working fluids to generate electricity from low temperature industrial waste heat. These heat pump cycles have been used to increase the efficiency of thermochemical cycles for hydrogen production from water. An Aspen Plus simulation has been developed to evaluate the effect of various working fluids on the net heat pump efficiency over a wide range of heat exchanger/evaporator output temperatures between 50°C and 250°C. There are 45 refrigerants as working fluids under investigation from the environmental classifications of HydroChloroFluoroCarbons (HCFC), HydroFluoroCarbons (HFC), ChloroFluoroCarbons (CFC) and HydroCarbons (HC) with boiling points between -88.65°C and 110.65°C. The simulation results reveal that the net efficiency increases with the increase of the inlet temperature of the waste heat source except with the azeotropic blends, R503 and R508B, since both are very low temperature refrigerants. The net efficiency obtained from this work ranges from 2% to 46%. Some of the results of the present study demonstrate that the working fluid R503 delivers the maximum efficiency among the evaluated working fluids for temperature <100°C; the zeotropic blend R410A shows the best efficiency for temperature between 100°C and 150°C; and for the temperatures between 150°C and 250°C, the azeotropic blend R500 exhibits the best efficiency. The highest net efficiency has been achieved with R503 at 50°C by varying the azeotropic composition of R23:R13 to 85:15 weight percent.

Key words: heat pump, working fluid, refrigerants, net efficiency, waste heat, thermochemical cycles, hydrogen production

See more of this Session: Nuclear Hydrogen Plant Design, Analysis and Commercialization
See more of this Group/Topical: Topical 8: Hydrogen Production for a Hydrogen Economy

Hydrogen Production by Sulphur Iodine Cycle Fed by Solar Energy: Realization of a Laboratory Plant and Possible Spin-off On the Industrial Field

Raffaele Liberatore, ENEA, Rome, Italy

Giampaolo Caputo, ENEA, Rome, Italy

Paolo Favuzza, ENEA, Rome, Italy

Claudio Felici, ENEA, Rome, Italy

Alberto Giaconia, ENEA, Rome, Italy

Michela Lanchi, ENEA, Rome, Italy

Giovanni Salvatore Sau, ENEA, Rome, Italy

Annarita Spadoni, ENEA, Rome, Italy

Pietro Tarquini, ENEA, Rome, Italy

In a desirable future in which energy could be not related to greenhouse emissions of gases into the atmosphere and without consuming resources owned from few countries, with its resulting strategic use, the hydrogen produced from water using renewable energy sources seems the best solution.

In this framework, ENEA has started a study of the thermolysis through thermochemical cycles powered by solar energy. The TEPSI project, in this regard, has as main objective, the construction of a plant for the production of 10 NL/h of hydrogen by sulfur-iodine cycle, which starting from water through a closed cycle of reactions that generate and consume iodine, hydriodic acid, sulfuric acid and sulfur dioxide, produces hydrogen and oxygen at a temperature compatible with current solar collectors technology and with a higher efficiency then the traditional methods of water thermolysis. A massive hydrogen production by this researched innovative methods will presumably be achieved by 2030, when all the equipments, built by appropriate materials, will be tested at pilot scale and when the solar technology has reached maturity levels allowing a suitable reduction of its impact on total production costs, which currently is about 2/3, but for that date is expected to fall by half. In the short term, however, it is possible using the knowledge learned in this study for purposes related to the industrial sector, such as the treatment of sulfur compounds from industrial processes. This method, patented by ENEA as part of this project, allows this result, at a cost-effective, without catalyst and with a contemporary production of hydrogen and sulfuric acid.

This research work concerns also the tests of the equipments related to this cycle, executed in the ENEA laboratories, with particular emphasis on the ones concerning the hydriodic section. These equipments, indeed, consume the most part of energy and, if bad designed, can invalidate the efficiency of the all cycle.

In the ENEA laboratory plant schema, the light phase of the Bunsen reaction ($2H_2O + I_2 + SO_2$ "³ $H_2SO_4 + 2HI$) composed of H_2SO_4 and H_2O is sent to an H_2SO_4 decomposer, after concentration in a distillation column, to generate SO_2, H_2O and O_2, which return to Bunsen reactor after a washing step to take away the unreacted species. The heavy phase, instead, is composed by a mixture containing HI, H_2O, traces of H_2SO_4 and a lot amount of I_2 (84-92 %wt). This phase, after purification from the H_2SO_4, is sent to a distillation column able to produce an azeotropic mixture of HI-H_2O (57 %wt of HI) as distillate and I_2 (95 %wt) as residue, which is sent to Bunsen reactor. By the carried out tests, the correct working of the column depends on the content of HI in I_2 free basis. If it is too low (< 55%wt) the column will produce a mixture containing a great amount of water and the separation of I_2 is not total, so different tests have been executed in order to find the best composition of the heavy Bunsen phase. The HI azeotropic mixture is sent to decomposer that works at 500 ¢XC and produces hydrogen and iodine with a conversion of about 22 %. The vapor from this equipment is condensed to separate hydrogen, but the liquid phase has an HI content of ~50%wt in iodine free basis and a content of iodine of ~11%wt. The I_2 purification of this stream is executed by the described column, to which is sent as lateral stream after to have taken away the water in excess in another packing column. This water obtained as distillate has resulted with very low iodine and HI content (< 0.5 %wt) and it is sent to Bunsen. The residue for the I_2 purification column is composed by a mixture of 14 %wt and 56.5 %wt of HI in iodine free basis.

This work analyses also the possible spin-off in the industrial field concerning the sulfuric compounds and the advantages in economical terms that are possible with the use of this cycle, using only tested equipments.

See more of this Session: Nuclear Hydrogen Plant Design, Analysis and Commercialization
See more of this Group/Topical: Topical 8: Hydrogen Production for a Hydrogen Economy

Experimental Corrosion Tests of Metals, Metal Alloys, and Ceramic Materials, in Boiling and High Temperatures Vapours I2-HI-H2O Mixtures for Sulphur-Iodine Thermochemical Cycle

Claudio Felici, ENEA, Rome, Italy

Paolo Favuzza, ENEA, Rome, Italy

Raffaele Liberatore, ENEA, Rome, Italy

Pietro Tarquini, ENEA, Rome, Italy

Adelina Borruto, Materials and Environment for Chemical Engineering, University of Rome "la Sapienza", Rome, Italy

Giovanni Narducci, Materials and Environment for Chemical Engineering, University of Rome "la Sapienza", Rome, Italy

The Sulphur-Iodine thermochemical cycle is one of the most promising way to produce hydrogen by splitting water. This cycle has been investigated by ENEA over the last five years, focusing on chemical aspects, reactions, total efficiency, technical feasibility and cost analysis. The majors goals for futures applications are:

- providing the energy demand with free carbon energy;

- rising as much as possible the total efficiency;

- solving some technical matter.

In order to realize this cycle in a massive way it is very important to find the materials for the equipments able to withstand the corrosive environment, which involves strong acids and oxidants in extreme conditions such as high temperatures and pressures. Only a few and incomplete data are available in literature on materials corrosion resistance in hydriodic acid (HI) environment. To overcome this lack, at the ENEA laboratories some corrosion tests are currently under investigation. Corrosion tests were set up in different conditions, in particular: boiling azeotropic HI (127 °C, 1 bar) and vaporized azeotropic HI at 500 °C, 1 bar, with the aim to simulate some critical components of a designed future plant, especially regarding the distillation columns and the HI decomposer.

The first apparatus consists in a heated 250 mL pyrex flask, where the samples have been dipped for 100 hours in a bath of boiling azeotropic HI-H2O and I2, above which a serpentine condenser is installed to avoid the vapour loss and to maintain the boiling equilibrium condition. The second device basically consists in a quartz tube (6 cm as diameter and 90 cm as length), placed in a tubular furnace at 500 °C. Inside the quartz tube a multi sample quartz holder is positioned. The azeotropic solution of HI has been pumped and injected inside the tube by a metering pump. Some quartz wool has been inserted in the initial part of the reactor in order to ensure the homogeneous vaporization and heating of the feed. Also, a little flow of Argon is sent inside the tube to guarantee a continuous and constant renewal of the vaporized mixture. Since the flows of reagents and carrier gas have been set to very low values, the test conditions can be considered quasi-static.

The vapour stream that comes out from the tube is forwarded in a flask, where it has been condensed and analysed by titration.

The tests concerned materials, such as:

Ti, Ta, Nb;

Ti-6Al-4V, Nb/Mo alloy and a large number of Ni-base Superalloys, provided by different companies;

Al_2O_3 coatings and Ceramic ;

Peek Carbon Fiber (Polyetherketone) and Torlon® Polyamide-imide (PAI).

After 100 hours of exposing time, the samples have been analyzed by the following techniques:

- Optical Microscopy (O. M.);
- Scanning Electron Microscopy (SEM);
- Energy Dispersive X-ray Spectroscopy (EDS);

Interesting conclusions have been found about the behavior of the materials, in these extreme conditions, because it has been possible to correlate their performances to the composition.

Moreover, the same materials showed a very dissimilar resistance at the variation of the environmental conditions. Generally, the higher was the temperature, the more aggressive resulted the environment, considering also that at high temperature some hydrogen has been produced inside the tube by the HI decomposition.

See more of this Session: High Temperature Materials and Systems for Hydrogen Production
See more of this Group/Topical: Topical 8: Hydrogen Production for a Hydrogen Economy

Application of Radiation Probes
to In Situ Composition Measurements of Bunsen Reaction Solution

Shinji Kubo[1], *Yasunobu Nagaya*[2]

[1]*Japan Atomic Energy Agency*
Narita-cho 4002, Oarai-machi, Higashi Ibaraki-gun, Ibaraki-ken, Japan

[2]*Japan Atomic Energy Agency*
Shirakata Shirane 2-4, Tokai-mura, Naka-gun, Ibaraki-ken, Japan

ABSTRACT

Concerning stable plant operations of the iodine-sulfur thermochemical water-splitting process, composition measurements of Bunsen reaction solution is one of key techniques. Radiation probes have a potential for applications to determine this multiple component solution while the non-contact approach avoids corrosive issues. Laboratory-scale experiments to examine its applicability were conducted with use of simulated HI-I_2-H_2O solution by a neutron moisture meter with a sealed radiation source. The obtained counting rates were correlated with hydrogen volume concentrations; this fact suggests that the neutron ray is a candidate probe to determine the composition of HI-I_2-H_2O solution.

INTRODUCTION

Measurement equipments are indispensable to chemical process controls. A stable hydrogen production by the iodine-sulfur thermochemical water-splitting process (IS process) is relatively difficult because of lack of existing in situ composition measurement techniques for multiple components and corrosive solution. Figure 1 shows a brief reaction scheme for IS process; Bunsen reaction produces two types of corrosive and multicomponent acids around 100 °C, a H_2SO_4 rich solution and a HI rich solution, which are separated by two-phase separation phenomenon. The HI rich solution is purified and concentrated, then decomposed. Products of I_2 should be recycled to Bunsen section. The H_2SO_4 rich solution is also processed in this manner. Looking at the return streams, closed cycle system causes that the streams concentrate on Bunsen section. Many streams with different composition return, so that Bunsen solution becomes changeable composition; accordingly, solution compositions which are consequent on Bunsen reaction, such as the HI rich solution, become also changeable.

A composition of a purified and concentrated HI rich solution—this solution consists of HI, I_2, and H_2O, which is called HIx solution—is an important index for stable process control, since a throughput of the HIx solution should be adjusted with the assistance of this composition value depending on the targeted hydrogen production rate. Methods by radiation probes(1) have a potential for applications to determine this multiple component solution while the non-contact approach avoids the corrosive issues. Moreover these probes have features of the promptness, contact-less, and continuous use. The purpose of this study is to evaluate applicability of the methods for measuring the composition of HIx solution; we examined data which was obtained by a neutron moisture meter through laboratory-scale experiments with use of simulated HIx solution.

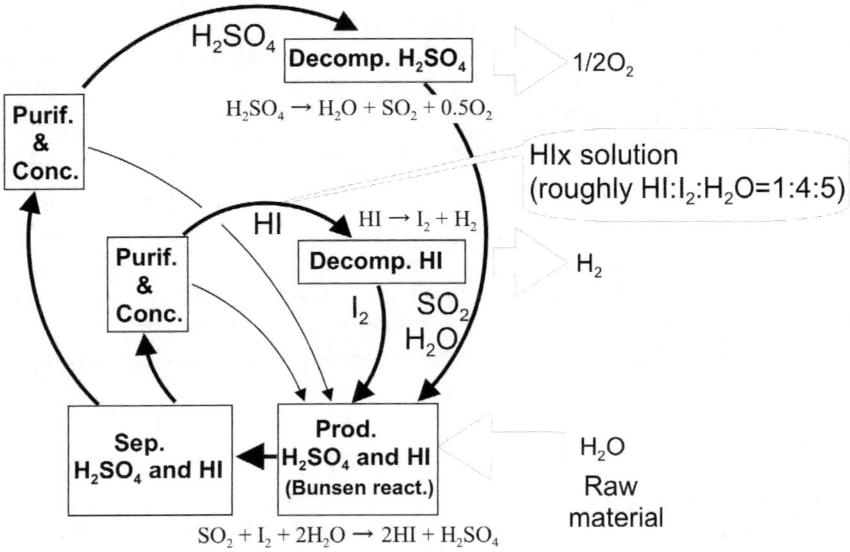

Figure 1: Reaction scheme of IS process. Solutions come out of Bunsen reaction are corrosive and multicomponent. Flow rate of HIx solution must be adjusted with assistance of composition of HIx solution, which must be obtained by measurement methods for stable plant operation.

IN SITU COMPOSITION MEASUREMENT BY RADIATION PROBES

Figure 2 shows a concept scheme of an in situ composition measurement by radiation probes which are gamma probes and a neutron probe. Two different gamma-ray sources in energy levels are employed to measure transmittance through a solution and a fast-neutron source

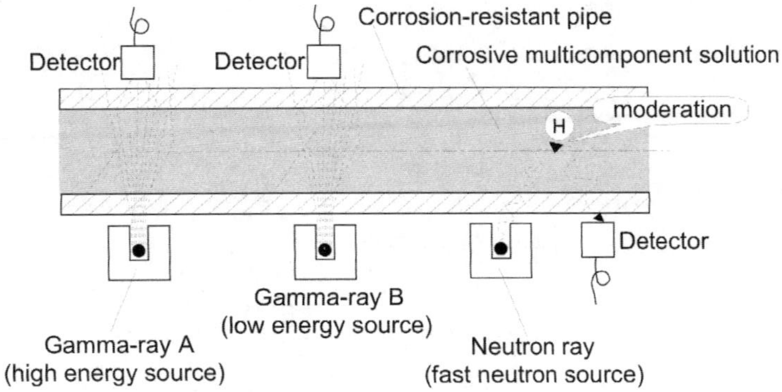

Figure 2: Conceptual scheme of in situ method in actual use to determine composition of corrosive and multicomponent solution. Non-contact approach avoids the corrosive issues. Two or three probes can be used properly depending on a type of a solution.

emitting fast neutrons is employed to measure thermal neutrons moderated by interactions with hydrogen atoms. Integrations of data from three different probes have possibilities that compositions of four component, at a maximum, solutions are determined. Two or three probes can be used properly depending on a type of a solution.

To determine the composition of HIx solution requires the two radiation probes, since the solution consists of three components of HI, I_2, and H_2O. It has been generally recognized that the correspondences between the gamma-ray transmittance and solution densities are strong; moreover, there is a direct correlation between the densities of HIx solution and the iodine concentrations(2). Accordingly, we focused on the hydrogen in HIx solution to explore residual or alternative information to complete the composition determination technique by the radiation probes. A system of neutron moisture meter employing the neutron ray provides advantages in quantification of the hydrogen.

EXPERIMENTAL

A schematic of experimental apparatuses is shown in Figure 3. A container, which is open to the atmosphere, made of a plastic with diameter of 0.22 m was placed on a top board made of a stainless steel. A sensor assembly of a neutron moisture meter (NMA-2001 ,SHINKO EN&M) was put on to the underside of the top board; a fast-neutron source (californium, 2.5 MeV) and helium-3 proportional counters were built into this assembly, which were sealed by polyethylene walls. The thermal neutrons moderated by the interactions with the hydrogen atoms were counted by the counters housed in the assembly. Densities and compositions of the test solutions were determined by a oscillating density meter and a chemical titrator. Temperature of the solutions were obtained by a thermocouple. The test solutions were prepared by mixing regents of iodine powder and hydriodic acid. It is possible to take test solution in and out of the container by a tube pump. Levels of the solution in the container were calculated with weight

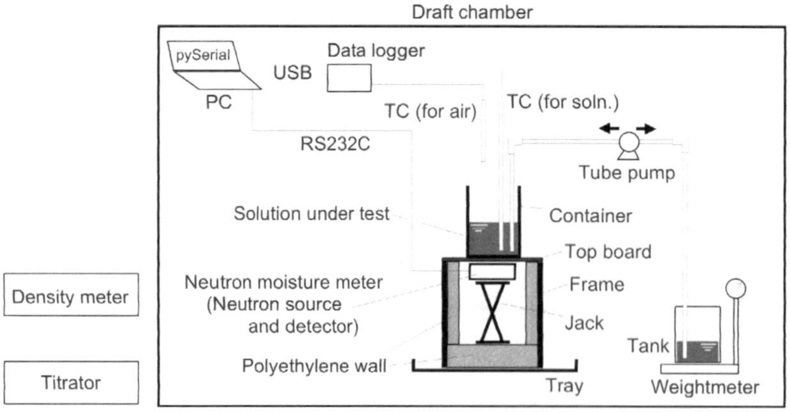

Figure 3: Experimental apparatus employing neutron moisture meter using fast neutron source to examine test solutions in container. Thermal neutrons moderated by interactions with hydrogen atoms were counted by detector. Densities and compositions of test solutions were determined by oscillating density meter and chemical titrator.

changes of the tank, the densities of the test solutions, and the diameter of the container.

RESULTS AND DISCUSSIONS

Figure 4 shows the counting rates as a function of water level. The counting rate at that the solution level equivalent to zero displays a significant value, since the apparatus materials itself affect the measured values. An effect on the counting rates against the solution level disappears as certain value of the solution level is exceeded. Accordingly, influences on the counting rates by difference of the solutions might be compared through the use of a difference between two values which are the counting rate at solution level of zero and that in the flat region. A result for a HI solution (without iodine) and two HIx solutions appears in Figure 5. The volume concentration of hydrogen atom were calculated from measured densities and titrated composition. This figure illustrates the counting rates are correlated with the volume concentration; this fact suggests that the neutron probe has a possibility to determine the composition of HIx solution.

In order to develop a prediction method which enable to compute the counting rates for the measuring system, a Monte Carlo method was applied to this composition measuring system. For preliminary computation, a difference of the detector response between water and HIx solution was obtained, which suggests possibilities that the detector responses for HIx solution by the radiation probes are predictable.

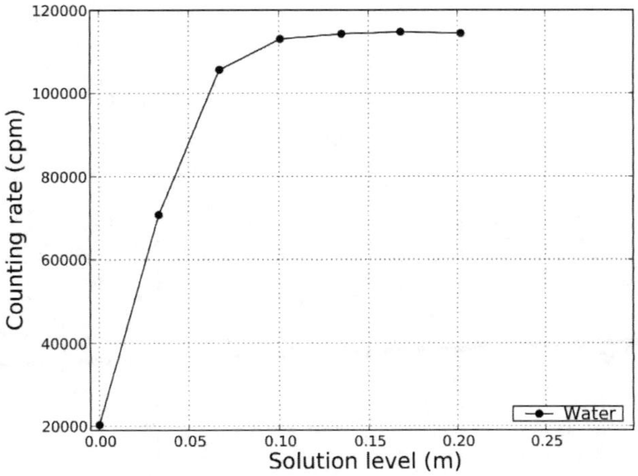

Figure 4: Counting rates of water by neutron moisture meter (23.5°C, ρ=0.997 g/cm^3). Counting rate at solution level of zero displays significant value. Counting rates in flat region must be evaluated to highlight differences of solution behaviors.

SUMMARY

The Measurement equipments are indispensable to chemical process controls. A stable hydrogen production by the Iodine-Sulfur thermochemical water-splitting process is relatively difficult because of the lack of existing in situ composition measurement techniques for multiple components and corrosive solution. Methods by radiation probes have a potential to determine the composition of this multiple component solution while the non-contact approach avoids the corrosive issues. We focused on the hydrogen in HIx solution to explore the technique to determine the composition for HIx solution. Laboratory-scale experiments to examine its applicability were conducted with use of simulated HI-I_2-H_2O solution by the neutron moisture meter with the sealed radiation source. The obtained counting rates were correlated with hydrogen volume concentrations; this fact suggests that the neutron ray is a candidate probe to determine the composition of HIx solution. Moreover, the application of the Monte Carlo method suggests possibilities that the detector responses for HIx solution by the radiation probes are predictable.

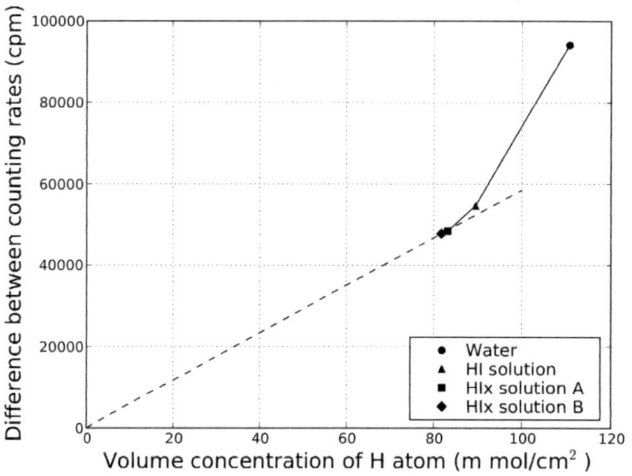

Figure 5: Counting rates of HI solution and HIx solution by neutron moisture meter (HI solution, 23.5°C, ρ=1.691 g/cm^3, $x_{\mathrm{HI}} : x_{\mathrm{I_2}} : x_{\mathrm{H_2O}} = 0.154 : 0 : 0.846$; HIx solution A, 22.5°C, ρ=1.861 g/cm^3, $x_{\mathrm{HI}} : x_{\mathrm{I_2}} : x_{\mathrm{H_2O}} = 0.146 : 0.026 : 0.828$; HIx solution B, 21.5°C, ρ=1.881 g/cm^3, $x_{\mathrm{HI}} : x_{\mathrm{I_2}} : x_{\mathrm{H_2O}} = 0.090 : 0.057 : 0.853$). Values on vertical axis represent differences between counting rate at solution level of zero and that in flat region.

ACKNOWLEDGMENT

This work was supported by KAKENHI (20560716).

REFERENCES

1. Hisao SUZUKI, Studies of Neutron Moisture Gauges, The Iron and Steel Institute of Japan, 57 (1971) 1, pp.175-181.

2. Shinji Kubo, Hayato Nakajima, Yoshiyuki Imai, Seiji Kasahara and Kaoru Onuki, Control Techniques For Bunsen Reaction Solution To Regulate Process Condition, AIChE Annual Meeting, 357f, Nov. 2007, Salt Lake City, USA.

The Effects of Space Time upon the Supercritical Water Reformation of Glycerol

Jared Bouquet[1], Jonathan Wenzel[1], Jason Picou[1], Michael Stever[1], Sunggyu Lee[1], Tae-Hoon Lim[2], and Byung Gwon Lee[2]

1. *Missouri University of Science and Technology, Rolla, MO*
2. *Korea Institute of Science and Technology, Seoul, KOREA*

Abstract

Glycerol is a less desirable byproduct formed in the production of biodiesel via transesterification of triglycerides and presents a nontrivial issue in terms of developing other beneficial end uses. Supercritical water reformation of glycerol presents a unique non-catalytic means of converting unwanted glycerol into hydrogen for use in energy and transportation applications, hence increasing the amount of usable transportation fuel that may be produced from triglycerides in biodiesel production. An experimental study of the supercritical water reformation of glycerol was conducted using a tubular 0.4-L Haynes® Alloy 230 reactor at a constant pressure of 22.4 MPa. The effect of space time was evaluated for the non-catalytic reformation of glycerol in supercritical water. The space time was varied between 30 and 160 seconds with temperatures ranging from 500 to 600°C and water-to-glycerol molar ratios ranging from 13.5:1 to 24:1.

Introduction

Glycerol, also known as glycerin, is a chemical with a multitude of applications. At ambient temperatures glycerol is a clear and viscous substance, and due to its nontoxic and sweet taste, it is a commonly used food and cosmetics additive. Most natural glycerol is produced in the manufacture of soap and in the hydrolysis of fats to create fatty acids.[1] The transesterification of triglycerides, such as vegetable oil, also produces glycerol, in addition to biodiesel. Biodiesel is an alternative liquid transportation fuel that has the potential to readily displace its petroleum counterpart, since it can be used in conventional diesel engines with little to no modifications.[2] With the emergence of biodiesel as an alternative fuel, finding means to utilize the glycerol byproduct has become an area of interest. The reason for this is that more glycerol is being created than can be used commercially in conventional enduses.[3] Finding another outlet for this excess glycerol will increase market demand and make biodiesel plants more economically profitable. One potential use for this excess glycerol is to make hydrogen, another alternative fuel. Hydrogen may be produced from glycerol by a novel process, non-catalytic supercritical water reformation. In order to determine the effect of space time on the supercritical water reformation of glycerol a series of experiments were conducted at 22.4 MPa at temperatures between 500 and 600°C and water-to-glycerol molar ratios ranging from 13.5:1 to 24:1.

Chemical Reactions

There are many chemical reactions that occur during the supercritical water reformation of glycerol. The five principal reactions discussed in this paper are the pyrolysis reaction, methanation reaction, reformation reaction, the water gas shift reaction, and the Boudouard reaction. The pyrolysis reaction involves the dismantling of glycerol into smaller hydrocarbons, such as methane, ethylene, and ethane. The ethane produced can still be reformed to make hydrogen or pyrolyzed even further to produce methane. While this reaction does not consume hydrogen, it does competitively consume the reactant for reformation, limiting hydrogen production. Equation 1 below shows one possible pyrolysis reaction.

The methanation reaction is a reversible reaction that involves carbon oxides and hydrogen reacting to become methane and water. The forward reaction consumes hydrogen, and therefore is detrimental to the reformation process. The methanation reaction and pyrolysis reaction are the only sources of methane for the purposes of this paper.

$$CO + 3H_2 \rightleftarrows H_2O + CH_4 \qquad \textbf{(1)}$$

The reformation reaction is endothermic and involves the reaction of water with glycerol to produce hydrogen as well as carbon monoxide. The reformation of hydrocarbons is commonly performed with the aid of a noble metal catalyst. The experiments presented in this paper were performed non-catalytically, with no catalyst loaded into the reactor, except potential monolithic catalytic effects from the reactor wall made of high-nickel alloy of Haynes® 230.

The forward water gas shift reaction (WGS) is fourth major reaction in the supercritical water reformation of glycerol. The reaction is reversible in nature, and involves the combination of water and carbon monoxide to make hydrogen and carbon dioxide. The forward reaction is thermodynamically favored at temperatures of 815°C or below.[4] Assuming that all carbon monoxide produced in the reformation reaction subsequently undergoes in the WGS reaction, the overall reaction would yield seven moles of hydrogen per mole of glycerol. The overall reaction is shown below.

$$C_3H_5(OH)_3 + 3\,H_2O \rightarrow 7\,H_2 + 3CO_2 \qquad \textbf{(2)}$$

This reaction means the maximum amount of hydrogen production is seven moles per mole of glycerol fed. Another possible reaction that could be happening is the Boudouard reaction. The Boudouard reaction is a reversible reaction that involves two carbon monoxide molecules combining to make a carbon dioxide molecule and elemental carbon. The effect of the Boudouard reaction is limited at higher temperatures; the equilibrium constant favors carbon monoxide production at temperatures above 680°C.[4-8] This means that the Boudouard reaction will only influence the experimental runs with lower temperatures. Increasing amounts of carbon dioxide and lower carbon monoxide and carbon gasification values can be indicative of the Boudouard reaction.

$$2CO \rightleftarrows C + CO_2 \qquad \textbf{(3)}$$

Apparatus and Chemicals

The chemicals used for the experiments were deionized water and 99.7% pure glycerol from thechemistrystore.com. Glycerol above 80 wt-% in purity will absorb ambient water vapor from the air; therefore the bulk glycerol was diluted to 75.9 wt-%. The water and glycerol for the experiments were then mixed into solutions with concentrations between 17.55 wt-% and 27.45 wt-% glycerol. These solutions correspond to a water-to-glycerol molar ratio of 24:1 and 13.5:1 respectively. The physical properties of glycerol and water may be found in Table 1.

Table 1. Physical Properties of Glycerin and Water[9,10]

Property	Glycerol	Water
Molecular Weight (g/mol)	92.09	18.02
Density at 25 ˚C and 1 bar (g/cm^3)	1.26	0.99
Melting Point (˚C)	17	0
Boiling Point (˚C)	287	100
Critical Temperature (˚C)	577	647
Critical Pressure (atm)	74.02	217.7

The supercritical water reformation system used for these experiments consists of a water and fuel feed system, reactant preheater, Haynes® Alloy 230 reactor with a heater, liquid collection system, gas sampling system, and a data acquisition and control system. Figure 1 is a representative process flow diagram of the supercritical water reaction system. The Haynes® Alloy 230 reactor has an internal volume of 383 mL, and is an alloy of mostly nickel, chromium, tungsten, and molybdenum, along with other elements.[11] The stress-strain, creep, and rupture properties of Haynes® Alloy 230 allows the reactor to operate at a temperature of 800°C at a pressure of 36.2 MPa. The samples collected of the gaseous effluent at ambient conditions were analyzed using a HP 5890 Series A gas chromatograph with a thermal conductivity detector. The gas chromatograph is calibrated to detect hydrogen, nitrogen, carbon monoxide, methane, carbon dioxide, acetylene, ethylene and ethane.

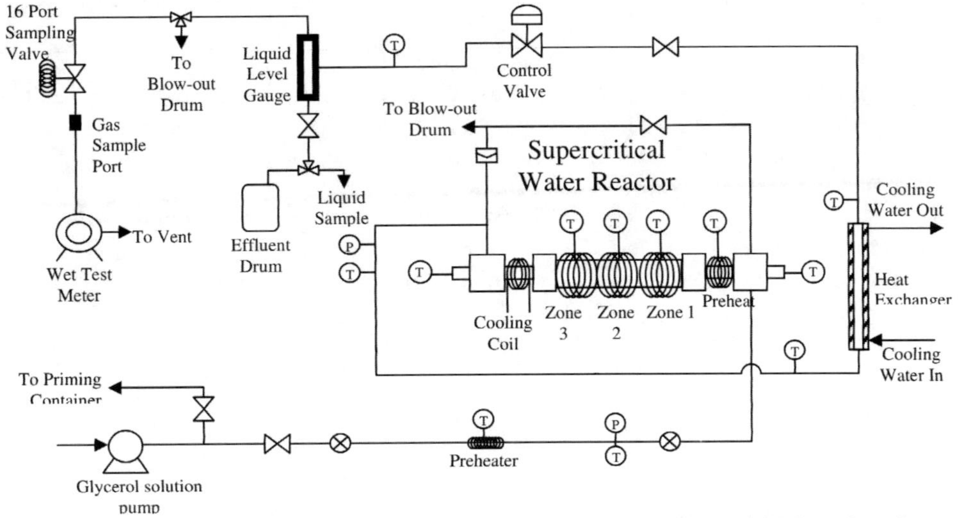

Figure 1. The supercritical water reformation system at the Missouri University of Science and Technology.

Experimental

The space time of the water-glycerol solution was varied for these experiments to determine the effect upon gas composition as well as the extent of the gasification of glycerol. The experiments were performed non-catalytically at a constant pressure of 24.13 ± 0.23 MPa. Space time was varied at temperatures of 500 ± 4°C and 600 ± 9°C and water to glycerol molar ratios of 13.5:1 and 24:1. The space time was determined using the Peng-Robinson equation of state with Van der Waals mixing rules. In order to hold the space time constant, the feed flow rate was adjusted to account for effects of experimental variables such as temperature. Table 2 lists the temperature, space time, pressure, and water/glycerol (W/G) ratio of each experimental run.

Table 2. Experimental Run Summary

Run No.	Temp (°C)	Space time (sec)	Pressure (psig)	W/G
WG-31	492	157	3507	13.5
WG-33	501	105	3515	13.5
WG-32	497	52	3506	13.5
WG-25	498	146	3497	24.0
WG-30	496	102	3511	24.0
WG-34	504	52	3499	24.0
WG-18	598	147	3491	24.0
WG-10	599	127	3503	24.0
WG-19	601	99	3481	24.0
WG-11	599	65	3505	24.0
WG-12	591	31	3503	24.0

Results and Discussion

The results of the experiments listed in Table 2. **Experimental Run Summary**Table 2 are shown below in Table 3 and include the yield of the different gaseous species leaving the reactor relative to the amount of glycerol fed, the total carbon conversion into gaseous species, and the yield of hydrogen gas produced compared to the theoretical maximum. Carbon not present in the gaseous effluent is either present in the liquid effluent, or remains in the reactor as a solid. The reactor was cleaned of any solid carbon residue either manually or by supercritical water oxidation prior to an experiment.

Table 3. Gas Effluent Flow Rates and Conversions Summary

Run No.	Gas Yield (mol gas/mol glycerin fed)							Carbon conversion	H_2 yield
	H_2	N_2/O_2	CO	CH_4	CO_2	C_2H_4	C_2H_6		
WG-31	0.76	0.00	0.65	0.19	0.33	0.02	0.05	44.8%	10.9%
WG-33	0.77	0.00	0.55	0.17	0.32	0.01	0.04	39.0%	11.1%
WG-32	0.54	0.00	0.52	0.09	0.09	0.02	0.02	26.6%	7.7%
WG-25	3.03	0.00	0.21	0.61	1.70	0.00	0.14	94.6%	43.5%
WG-30	2.79	0.00	0.26	0.55	1.50	0.00	0.11	85.7%	40.1%
WG-34	2.12	0.00	0.64	0.46	0.91	0.01	0.07	73.3%	30.5%
WG-18	3.10	0.00	0.17	0.80	1.75	0.00	0.09	98.2%	44.6%
WG-10	3.11	0.00	0.17	0.71	1.73	0.00	0.08	93.7%	44.8%
WG-19	3.43	0.00	0.24	0.69	1.82	0.00	0.10	100.0%	49.4%
WG-11	3.47	0.00	0.24	0.58	1.72	0.00	0.08	91.5%	49.9%
WG-12	3.44	0.00	0.37	0.53	1.62	0.00	0.08	90.9%	49.6%

The effect of space time upon the gas composition at various temperatures may be found in Figures 2 through 4. Each figure shows the effect of increasing space time on the composition of the gas effluent, as well as the total carbon and hydrogen yield.

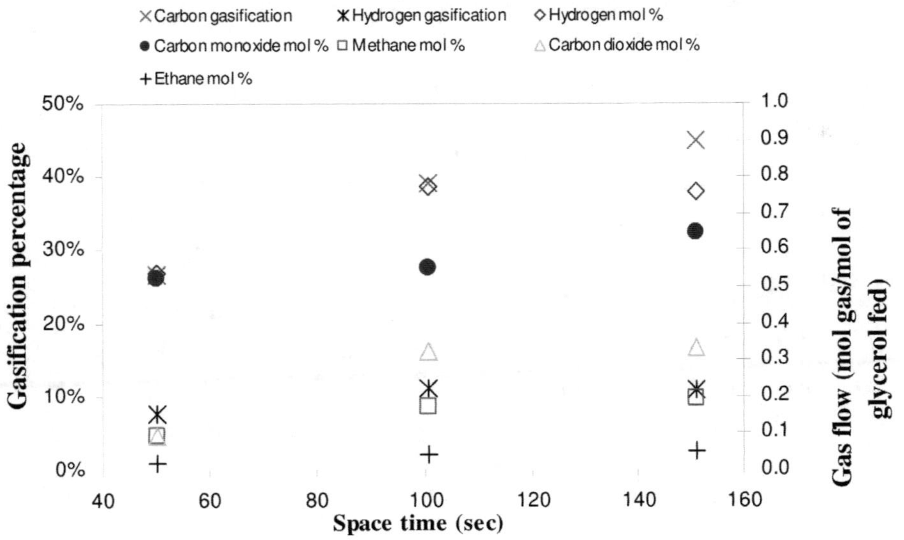

Figure 2. Effect of Space time at 500°C and 13.5 W/G

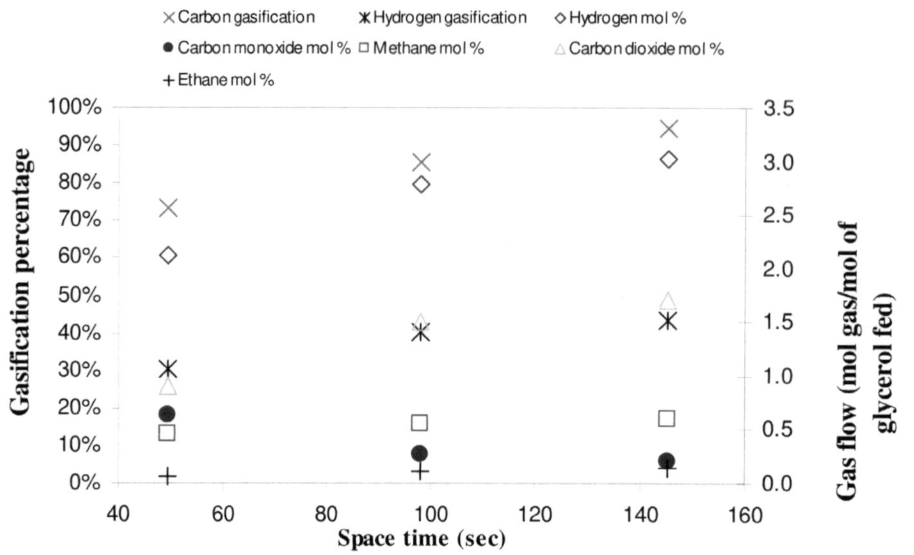

Figure 3. Effect of Space time at 500°C and 24 W/G

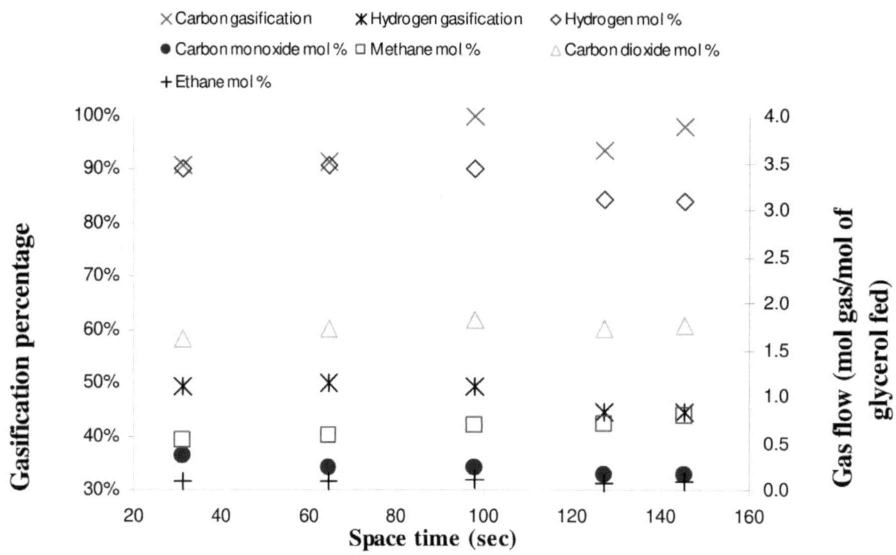

Figure 4. Effect of Space Time at 600°C and 24 W/G

All three experimental sets show an increase in carbon gasification with increasing space time. However, other than carbon gasification there is no other trend that is the same in all three experimental sets. At different temperatures and W/G ratios, different trends with respect to space time are observed.

At the relatively low temperature of 500°C and a 13.5 W/G ratio the carbon conversion is far from completely reaching a maximum of 44.8% at the highest space time. This means the balance of the carbon is remaining in the reactor or coming out in the liquid effluent. Figure 2 reveals an overall increase in the production of all gases with increasing space time. Figure 2 also shows that at the greatest space time the only species to show a significant increase from the previous space time is carbon monoxide with an increase of 0.1. Along with almost no increase in hydrogen and carbon dioxide between those space times, it could mean that the water gas shift reaction is not as reactive at those run conditions.

The values in Figure 3 show a greater overall carbon conversion at run conditions of 500°C of and a 24 W/G ratio. This can be attributed to the more dilute conditions, with a maximum carbon conversion of 94.6% at the greatest space time. The only difference in trends from Figure 2 to Figure 3 is a decrease rather than increase of carbon monoxide with respect to space time. Also observed was an increase in hydrogen and carbon dioxide values. These results may be attributed to an increase in the forward water gas shift reaction. Even though the low temperature, increase in carbon dioxide, and decrease in carbon monoxide may indicate the Boudouard reaction is occurring, the increase in carbon gasification and increasing hydrogen yield indicate water gas shift is more likely.

The third set of experimental runs at 600°C of and a 24 W/G ratio show nearly complete carbon conversion, with a conversion of 90.9% at the lowest space time. Figure 4 shows that hydrogen and carbon monoxide gas are decreasing with increasing space time. This trend may be the result of pyrolysis or methanation occurring with respect to increasing space time. This is supported by the fact that the only gas to show an increase in production with respect to increasing space time at these conditions is methane.

Conclusion

Overall, all three sets of experiments showed an increase in the gasification of glycerol with an increase in space time. The effects of space time on which reactions took priority for each of the three experimental run sets were shown. Since temperature and W/G ratio also influence the reaction rates, the effect of space time was different for each experimental set. At 500°C and a 13.5 W/G ratio, increasing space time showed a reduced effect of the water gas shift reaction. At the same temperature and a 24 W/G ratio, an increasing space time showed an increase in the forward water gas shift reaction. Increasing space time at 600°C and a 24 W/G ratio increased the amount of methanation and pyrolysis.

References

(1) Jungermann, Eric; Sonntag, Norman O.V. Glycerine, a Key Cosmetic Ingredient. Marcel Dekker Inc. New York, 1991; pp 3.
(2) Johnson, D.T.; Taconi, K.A. The Glycerin Glut: Options for the Value-Added Conversion of Crude Glycerol Resulting from Biodiesel Production. Published online Oct. 1, 2007 in Wiley InerScience
(3) Nilles, Dave. Combating the Glycerin Glut. *Biodiesel Magazine*.[online] **2006**, *25* Article Archive http://www.biodieselmagazine.com/article.jsp?article_id=1123&q=&page=all (accessed September 15, 2009)
(4) Lee, S. *Alternative Fuels*, Taylor & Francis, Philadelphia, **1996**; pp.140-151.

(5) McIntosh, S.; Gorte, R.J. Direct Hydrocarbon Solid Oxide Fuel Cells. *Chem. Rev.* **2004**, *104*, 4845-4865.
(6) Murray, E. P.; Tsai, T.; Barnett, S. A. Adirect-methane fuel cell with a ceria-based anode *Nature* **1999**, *400*, 649.
(7) Garcia-Serna, J.; Garcia-Merino, E.; Cocero, M.J. Gasification of charcoal using supercritical CO_2 at high pressure. *J. of Supercritical Fluids* **2007**, *43*, 228-235.
(8) Mondal, K.; Lorethova, H.; Hippo, E.; Wiltowski, T.; Lalvani, S.B. Reduction of iron oxide in carbon monoxide atmosphere-reaction controlled kinetics. *Fuel Processing Technology* **2004**, *86*, 33-47.
(9) Glycerin Phase Change Properties. National Institute of Standards and Technology. http://webbook.nist.gov/cgi/cbook.cgi?ID=C56815&Units=SI&Mask=4#Thermo-Phase
(10) Water Phase Change Properties. National Institute of Standards and Technology. http://webbook.nist.gov/cgi/cbook.cgi?ID=C7732185&Units=SI&Mask=4#Thermo-Phase
(11) *Haynes® 230 Alloy*, Haynes High-Temperature Alloys International, 2004 pp. 2.

Influence of pH On Synthesis of Homogeneous Cobalt-Doped Barium Cerium Zirconium Oxide Via Oxalate Co-Precipitation

Aravind Suresh, Chemical, Materials & Biomolecular Engineering and Connecticut Global Fuel Cell Center, University of Connecticut, Storrs, CT

Benjamin. A Wilhite, Chemical, Materials & Biomolecular Engineering and Connecticut Global Fuel Cell Center, University of Connecticut, Storrs, CT

Prabhakar Singh, Chemical, Materials & Biomolecular Engineering and Connecticut Global Fuel Cell Center, University of Connecticut, Storrs, CT

Objective of the research is to develop a homogeneous, high-temperature, bi-functional, electro-ceramic catalyst for simultaneous hydrogen generation from methanol and hydrogen purification by mixed protonic electronic conductivity. Aliovalent cation-doped barium cerates are known to display significant protonic conductivity at elevated temperatures under hydrogen containing atmospheres. Zirconium substitution is expected to improve chemical stability of the material and cobalt doping is expected to impart protonic and electronic conductivity to the material in addition to catalytic activity towards hydrogen generation reactions. Precursor for cobalt-doped barium cerium zirconium oxide was synthesized via the route of co-precipitation using metal nitrates as starting compounds and oxalic acid as precipitant. Final oxide is obtained by calcining the precursor at high temperature. Synthesis experiments were carried out under different pH values in aqueous medium and at a fixed temperature. Filtrate and precipitate samples were subjected to elemental analysis to understand the influence of reaction pH on the degree of precipitation of the four metals from solution.

See more of this Session: High Temperature Materials and Systems for Hydrogen Production
See more of this Group/Topical: Topical 8: Hydrogen Production for a Hydrogen Economy

°METHOD AND SYSTEM FOR THE SEPARATION OF A MIXTURE CONTAINING CARBON DIOXIDE, HYDROCARBON, AND HYDROGEN

Baruchi Barry Kimchi, Rishon LeZion, Israel

ABSTRACT

The invention provides a method and system for the separation of a mixture containing H_2, hydrocarbon, and CO_2. The generated mixture is introduced into a distillation column having a side stream to generate a top stream comprising H_2, a middle (volatility) stream comprising hydrocarbon; and a bottom stream comprising CO_2. In a preferred embodiment, the mixture further comprises N_2 which is obtained in the top stream. In a most preferred embodiment, the H_2 and the N_2 are present in a molar ratio of 3 H_2:1 N_2. The generated H_2 and N_2 may be used for the synthesis of ammonia. Thus, the invention also proves a method and system for the generation of ammonia. The method and system of the invention was implemented on the process simulator UniSim Design Version R370Build 13058 of Honeywell.

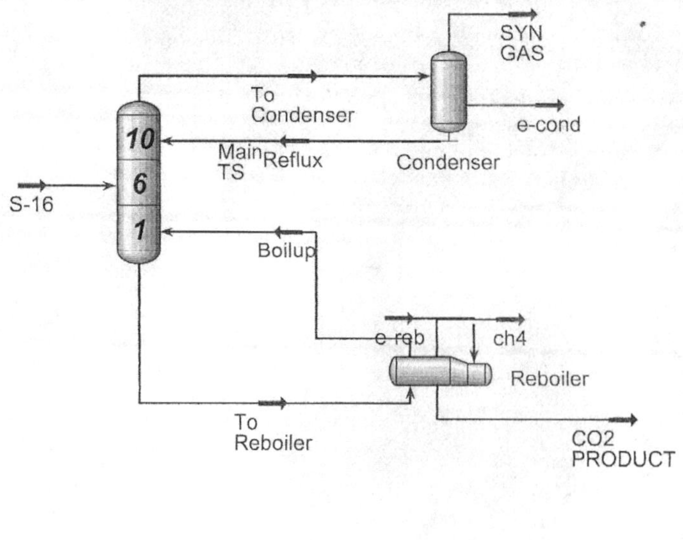

Fig1: Ammonia's Syn-Gas Separation System (Side stream withdrawn from the reboiler).

BACKGROUND OF THE INVENTION

The term *"synthesis gas"*, also known as *"syngas"* refers to a gas mixture containing carbon dioxide and/or monoxide and molecular hydrogen generated by the gasification of a carbon-containing fuel to a gaseous product with a heating value. Syngas is produced, for example, by steam reforming of natural gas or liquid hydrocarbons to produce hydrogen, the gasification of coal and in some types of waste-to-energy gasification facilities. Syngas is used, for example, as intermediates in creating synthetic natural gas, and for producing ammonia or methanol. Syngas is also used as an intermediate in producing synthetic petroleum for use as a fuel or lubricant.

In the synthesis of ammonia from stochiometric air (as a source of nitrogen) and hydrocarbons (as a source of hydrogen), the hydrocarbon, such as methane, is made to react with steam at elevated temperatures to generate H_2, CO, CO_2 H_2O. This produces a raw syngas containing, in addition to these compounds, residual unreacted hydrocarbon, as well as N_2, and other air constituents. The N_2 and H_2 must then be separated from the other components of the syngas for the generation of ammonia.

There are two main problems in the production of ammonia. One problem relates to the fact that excessive amounts of hydrocarbon, typically methane, remain unreacted in the conversion of the hydrocarbon to H_2 and CO, so that the ammonia yield is far from optimal. This so-called hydrocarbon slip can be reduced by using high reforming temperatures. Secondly, CO_2 must be removed from syngas to prevent poisoning of the catalyst used in the ammonia conversion, and this CO_2 removal requires high capital costs and is also costly in terms of energy consumption.

Significant work has been applied to the development of methods for the removal of carbon dioxide from a syngas. The processes can be separated into four general classes; absorption by physical solvents, absorption by chemical solvents, adsorption by solids, and distillation.

The high relative volatility of methane with respect to carbon dioxide makes cryogenic distillation theoretically very attractive. However, the methane/carbon dioxide distillative separation has a significant disadvantage in that solid carbon dioxide exists in equilibrium with vapor-liquid mixtures of carbon dioxide and methane at particular conditions of temperature, pressure, and composition. Obviously, the formation of solids in a distillation tower has the potential for plugging the tower and its associated equipment. Increasing the operating pressure of the tower will result in warmer operating temperatures and a consequent increase in the solubility of carbon dioxide, thus narrowing the range of conditions at which solid carbon dioxide forms. However, additional increases in pressure will cause the carbon dioxide-methane mixture to reach and surpass its critical conditions. Upon reaching criticality, the vapor and liquid phases of the mixture are indistinguishable from each other and therefore cannot be separated. A single-tower equilibrium separation operating in the vapor-liquid equilibrium region bounded between carbon dioxide freezeout conditions and the carbon dioxide-methane critical pressure line may produce a product methane stream containing 10% or more carbon dioxide.

Various methods have been devised to avoid the conditions at which carbon dioxide freezes and yet obtain an acceptable separation. Two processes which utilize additives to aid in the separation are disclosed in U.S. Pat. No. 4,149,864 to Eakman et al, and U.S. Pat. No. 4,318,723 to Holmes et al.

Eakman et al discloses a process for separating carbon dioxide from methane in a single distillation column. If insufficient hydrogen is present in the column feedstream, hydrogen is added to provide a concentration from about 6 to 34 mole percent, preferably from about 20 to about 30 mole percent. The separation is said to take place without the formation of solid carbon dioxide. The tower pressure is preferably held between 1025 and 1070 psia.

US Patent No. 4, 511, 382 to Valencia et al discloses separating acid gases, particularly carbon dioxide, from methane by cryogenic distillation in which an effective amount of a light gas, preferably helium, is added to a stream containing methane and carbon dioxide and cryogenically distilling the mixed stream to produce a liquid carbon dioxide stream and an enriched methane stream. The distillation tower or at least a portion thereof may then be operated at a pressure higher than the critical pressure of methane.

US Patent No. 7,090,816 to Malhotra et al discloses a method for the purification of syngas, such as occurs in the manufacture of ammonia, using cryogenic distillation. Refrigeration for the distillation is obtained from waste fluid expansion using a liquid expander to recover mechanical work from the waste fluid. This method reduces pressure loss in the syngas stream and reduces compression and power relative to similar ammonia generating processes.

SUMMARY OF THE INVENTION

In one of its aspects, the invention provides a method for the separation of a mixture containing H_2, hydrocarbon, and CO_2. In accordance with this aspect of the invention, the mixture is introduced into a distillation column having a side stream. Distillation of the mixture using a column having a side stream generates three streams, as follows:

(i) a top stream comprising H_2;

(ii) a middle stream comprising hydrocarbon; and

(iii) a bottom stream comprising CO_2.

The hydrocarbon of the system is preferably, although not necessarily, methane. In a preferred embodiment of the method and system of the invention, the mixture further comprises N_2. In a most preferred embodiment, the H_2 and N_2 are present in a molar ratio of 3:1. This may be achieved by mixing the hydrocarbon with a stoichiometric amount of air, as is known in the art. Embodiments in which the H_2 and N_2 are present in this molar ratio are useful for generating H_2 and N_2 for use in the manufacture of ammonia. Thus, in another of its aspects, the invention provides a method and system for the production of ammonia. In accordance with this aspect of the invention, a mixture containing H_2, N_2, hydrocarbon, and CO_2 is introduced into a distillation column to produce the three streams described above. The middle stream can be recycled and re-reformed. The top stream comprising H_2 and N_2 is then used to generate ammonia by any method known in the art.

The process has several degrees of freedom allowing flexibility in determining the operating parameters of the system, such as methane slip and side stream composition, and can improve the efficiency of the raw syngas generating process.

Liquid CO_2 which is obtained as the column's bottoms may be sequestrated and traded in the carbon market. Economical balance accounted for liquid CO_2 profits.

Experimental data for the five component mixture of this invention is unavailable in the literature. However, from an analysis of the phase diagram of the corresponding binary system (CO_2/CH_4), and the fact that a high concentration of H_2 leads to a significant <u>increase</u> of the critical pressure and also to a decrease in the freezing pressure of the CO_2. It can be concluded that for the "real-life" multi-component mixture at hand, pressure range for operation, in which CO2 freezing is prevented, may be wider than for the binary CO_2/CH_4 mixture. It is noted that the rest of the gasses of the multi-component mixture has a lower critical temperature than methane In addition, it's worth noting that for the multi-component mixture, vapor pressure line of pure CH_4 will not limit separation boundary at low temperatures.

For the limiting case of a CO_2/CH_4 mixture the following can be found:

a) CO_2 freezing is pressure and concentration dependent:
if every tray in the column would be kept above ~ 700 Psia only liquid/vapor equilibrium should exists (for the real life mixture this pressure may be lower)

b) The lower the CO_2 concentration is, the lower temperatures and working pressures are enabled.

c) Other light gasses - such as Hydrogen will significantly contribute to the success of the separation. (lowers operation pressure, separation is not bounded by the critical pressure of pure CH4)

"Syn-Gas" composition which is used on the Ammonia process (table 1) offers the opportunity to combine all of the above:

Table 1			
Feed Stream (a)	Flow rate [kgmol/hr] (b)	molar fraction – y (c)	partial pressure [atm] (d)
CO_2	654	0.164	2.455
H_2	2206	0.552	8.280
CH_4	368	0.092	1.381
N_2	751	0.188	2.819
Ar	17.59	0.004	0.066
Total	3996.59	1.000	15

Table 1 – example of a raw "syn-gas" composition used in Ammonia process

a) "Syn-Gas" is Hydrogen rich (~ 0.55 mol percent)

b) When the side stream is withdrawn a complete CO_2/CH_4 separation is not necessary , a side stream which contains ~ 20% CO_2 can be recycled - thus enabling working with higher column pressures or higher (in oppose to other processes eg LNG which requires less than 2% CO_2 and thus require a more complex separation)

c) CO_2 feed concentration can be lowered by feeding it directly into the reboiler of the column, withdrawing liquid CO_2 or by flushing 85% of it prior entering to the column.

The method and system of the invention was implemented on the process simulator UniSim Design Version R370Build 13058 of Honeywell. The thermodynamic package used was the Peng Robinson, Peng Robinson Sour Vapor packages.

Fig. 1 shows a system for separating a mixture containing H_2, hydrocarbon, and CO_2, in accordance with one embodiment of the invention.

DETAILED DESCRIPTION OF EMBODIMENTS

Fig.1 shows a system for separating a feed mixture containing H_2, N_2, hydrocarbon and CO_2. The system comprises a distillation column, a reboiler and a condenser. The distillation column comprises one or more trays, as is known in the art for distillation columns. The feed mixture may be a gas, or liquid/gas mixture. The feed mixture is introduced into the distillation column at an inlet and is deposited into a feed tray of the column. After the mixture reaches equilibrium in the feed tray, the H_2 and N_2, flow up the column while the CO_2 is liquefied and flows to the bottom by gravity. The working pressure inside the column should be above about 5 bar in order to assure that the CO_2 can exist in the liquid phase in the column. Part of the hydrocarbon entering the column is liquefied, mixed with the CO_2, and then flows down towards the bottom of the column. Gaseous hydrocarbon first rises in the column and liquefies by the liquid nitrogen, and then flows down and re-evaporates. In every tray there is a liquid/vapor equilibrium whose composition is determined by the tray's temperature.

The distillation column generates a condenser feed stream containing primarily the H_2 and N_2 in gaseous form. The output stream is introduced into a condenser that generates a liquid reflux that returns to the column, preferably to the top tray, of the column. The reflux of the top stream to the column is preferably performed using a reflux ratio between 0.001 and 10, and more preferably between 0.5 and 2. It is possible to alter the temperature gradient in the column by varying the reflux rate.

The reflux serves as a cooling source inside the column for the trays above the feed tray. In the lower part of the column, where the temperature is higher, only small amounts of liquid nitrogen are present, and most of the cooling for the trays below the feed tray is provided by liquid hydrocarbon and liquid CO_2.

An overhead product stream containing primarily gaseous H_2 and N_2 is drawn off from the condenser. As explained above, the overhead product stream can be used in the synthesis of ammonia. The distillation column also generates a reboiler feed stream containing primarily hydrocarbon and CO_2. The reboiler feed stream is introduced into the reboiler. In the reboiler, hydrocarbon boils, and may be withdrawn as a vapor side stream, while liquid CO_2 is withdrawn as a bottom stream. The liquid CO_2 in the bottom stream is easier to dispose of than gaseous CO_2. The reboiler generates a boilup that is returned to the column preferably to the bottom tray of the column. The vapor side stream can be recycled and reformed and used to generate new feed stream. Recycling of the hydrocarbon increases the utilization of the hydrocarbon, thus increasing the ammonia yield.

The column preferably has a pressure between 5 bar to a critical pressure of the mixture, and more preferably between 7 bar to 55 bar.

Table 2 shows the energy consumption. Energy costs were calculated assuming a use of a nitrogen refrigerant utility at a cost of 1 million dollar/ million kcal/hr/yr. The energy demand and cost were determined by the reflux ratio in this example the reflux ratio is 1.9. When running the column at 15 bar the reflux composition was primarily liquid nitrogen.

Table 2			
Condenser duty	$7.08X10^6$ kcal/hr	Energy cost of Condenser	$7.08 million/year
Reboiler duty	$1.06610X^6$ kcal/hr	Energy cost of Reboiler	$1.0661 million/year

Table 2: Column's Energy consumption (Reflux Ration = 1.9)

In an alternative embodiment, (not shown), a side stream is withdrawn from a tray in the column, instead of withdrawing the side stream from the reboiler. In another embodiment, the feed stream is introduced directly into the reboiler which is set to conditions under which CO_2 is a liquid at its bubble point and is withdrawn.

Specie	Bp at 1 atm (Celsius)
Hydrogen	-252
Nitrogen	-195.8
Argon	-185.7
Methane	-161
CO2	-57

Table 3: Boiling points of column's species:

Sulfur-Iodine Cycle: Phase Equilibrium Data for the Ternary Iodine-Water-HI and the Binary Iodine-Water Systems

Sarah E. Mena, Chemical and Biomolecular Engineering, Clemson University, Clemson, SC

Eduardo Cervo, Chemical and Biomolecular Engineering, Clemson University, Clemson, SC

David A. Bruce, Chemical and Biomolecular Engineering, Clemson University, Clemson, SC

Mark Thies, Chemical and Biomolecular Engineering, Clemson University, Clemson, SC

One of the primary targets of international interest for the centralized production of hydrogen from nuclear power is the Sulfur-Iodine cycle, which involves highly non-ideal phase behavior between iodine, water, and hydroiodic acid (HI) at elevated temperatures and pressures. Previous studies have revealed the need for experimental data in order to construct more reliable physical property models for Section III of the cycle, which involves the decomposition of HI into molecular hydrogen and iodine.

To this end, a continuous-flow apparatus (CFA) was constructed to measure the phase equilibrium for the binary iodine-water and the ternary water-iodine-HI systems. The highly corrosive nature of the system requires special attention. Several versions of the apparatus incorporating different materials, including polymers such as Teflon and PEEK and alloys such as Hastelloy C and tantalum-tungsten Ta-W, were made. Of these, only Ta-W was found to successfully withstand the highly corrosive environment at the temperatures and pressures of interest; in addition PEEK was found to be suitable for ambient-temperature application. Ta-W components, such as valves, fittings, and the equilibrium view cell, were fabricated at Clemson University.

The CFA consists of an arrangement of two separate streams, one of compressed, hot, molten iodine and one of a homogeneous solution of water and HI, that flow to an isothermal bath, where they are combined and brought to equilibrium in an impingement mixer/settler before entering the view cell. Here the phases split gravimetrically into a water-rich phase (light phase) and an iodine-rich phase (heavy phase) and exit through ports located at the top and bottom of the cell. Because the iodine and water-rich phases are essentially opaque to the naked eye, an infrared lens is used to locate the fluid-fluid interface in the view cell, either vapor-liquid (V-L), liquid-liquid (L-L), or even liquid-liquid-vapor. The flow rate of the exiting streams and system pressure are controlled using micrometering valves, with the exiting streams being collected in order to determine phase compositions.

The CFA allows us to obtain LLE, VLE, and LLVE for the Sulfur-Iodine cycle over a wide range of temperatures and pressures. For instance, LLE for the binary iodine-water has been observed at temperatures from 150 to 300 °C. These results are the first known observations of LLE for this system at temperatures above 225 °C. LLE for the ternary iodine-water-HI was also observed, and the boundary of the two-phase envelope for the water-rich phase has been mapped out for temperatures of 160 and 200 ºC, with higher temperature measurements up to 300 °C being planned. Measurement of equilibrium tie lines for the water-rich and iodine-rich phases is also underway and will be reported.

See more of this Session: Advances in Thermochemical Hydrogen Production
See more of this Group/Topical: Topical 8: Hydrogen Production for a Hydrogen Economy

Prevention of Sulfur Formation in Hybrid Sulfur Electrolyzer

John L. Steimke, Savannah River National Laboratory, Aiken, SC

Timothy J. Steeper, Savannah River National Laboratory, Aiken, SC

David T. Herman, Savannah River National Laboratory, Aiken, SC

The Hybrid Sulfur Cycle for producing hydrogen uses an electrolyzer with a sulfur dioxide depolarized anode. The cell voltage is much lower than for conventional electrolysis, however during testing of a Hybrid Sulfur electrolyzer a layer of sulfur frequently formed between the cathode and the membrane of the membrane electrode assembly in the electrolyzer. This layer increased cell voltage and eventually damaged the cell. Savannah River National Laboratory has developed an operating method to prevent sulfur formation. This method has been successfully tested over periods as long as ten days. To allow extended testing 24 hours a day the test facility has been heavily automated.

See more of this Session: Advances in Thermochemical Hydrogen Production
See more of this Group/Topical: Topical 8: Hydrogen Production for a Hydrogen Economy

Effect of Water Transport On the Hybrid Sulfur Electrolyzer

John Staser, University of South Carolina, Columbia, SC

John Weidner, Chemical Engineering Department, University of South Carolina, Columbia, SC

Efficient production of clean hydrogen on a large scale is necessary for the transition to a hydrogen economy. Traditional methods of hydrogen production use fossil fuels, which are not renewable and contribute to greenhouse gas emissions. In addition, the hydrogen produced by traditional methods may contain impurities. The hybrid sulfur thermochemical cycle, because it does not require fossil fuels, can potentially produce large quantities of hydrogen without contributing to greenhouse gas emissions. Its high-temperature step makes this process ideally suited to be used with advanced nuclear reactors. Researchers at the University of South Carolina have identified the critical design and operating parameters of the electrolysis step of the hybrid sulfur cycle. Realizing that water transport in the polymer electrolyte membrane is critical to electrolyzer operation, a mathematical model has been developed and coupled with experimental data to identify the key components of water transport. Here we present the water transport model, and discuss the effects of water transport on electrolyzer operation via its influence on reversible voltage and sulfuric acid concentration.

See more of this Session: Advances in Thermochemical Hydrogen Production
See more of this Group/Topical: Topical 8: Hydrogen Production for a Hydrogen Economy

Hybrid Sulfur Cycle Development for the Next Generation Nuclear Plant

William A. Summers, Hydrogen Technology Center, Savannah River National Laboratory, Aiken, SC

Maximilian B. Gorensek, Computational Science and Statistics Department, Savannah River National Laboratory, Aiken, SC

David T. Hobbs, Seperations Science Programs, Savannah River National Laboratory, Aiken, SC

John L. Steimke, Savannah River National Laboratory, Aiken, SC

Hector R. Colon-Mercado, Savannah River National Laboratory, Aiken, SC

Timothy J. Steeper, Savannah River National Laboratory, Aiken, SC

David T. Herman, Savannah River National Laboratory, Aiken, SC

This presentation will describe the latest developments for the Hybrid Sulfur cycle, which is being considered for demonstration with the Next Generation Nuclear Plant. It will include a status update on the SO2-depolarized electrolyzer (present vs. projected performance, scale-up, component development, etc.) as well as performance projections for the pilot and engineering scale plants.

See more of this Session: Advances in Thermochemical Hydrogen Production
See more of this Group/Topical: Topical 8: Hydrogen Production for a Hydrogen Economy

A Proposed New Sulfur-Sulfur Thermochemical Cycle

Alex Yokochi, School of Chemical, Biological and Environmental Engineering, Oregon State University, Corvallis, OR

Nicholas AuYeung, School of Chemical, Biological and Environmental Engineering, Oregon State University, Corvallis, OR

Of the many thermochemical cycles proposed for the production of hydrogen from water, the Sulfur-Iodine cycle is regarded as highly promising. In order to enable large scale implementation of this cycle, approaches that simplify product stream management following the Bunsen reaction are particularly important, especially with regards to the management of excess water and excess iodine. Excess water requires energy-intensive distillation schemes for the concentration of both H_2SO_4 and HI, the latter of which forms an azeotrope in water and requires reactive distillation.[1] Further, the global supply of iodine would need to grow by orders of magnitude in order for widespread implementation of the sulfur-iodine cycle.[2] In order to attempt to circumvent some of these problems, we have started to investigate possible process advantages derived from carrying out the Bunsen reaction in an ionic liquid as a low vapor pressure medium.[3] Beyond examining the kinetics of the Bunsen reaction itself and the evolution of Hydrogen Iodide and Sulfuric Acid from the reaction mixture, we identified conditions under which the reaction system may be biased to intentionally produce a stream of hydrogen sulfide based on a normally undesired side-reaction: [4]

$$8HI + H_2SO_4 = H_2S + 4H_2O + 4I_2$$

Since the weakly acidic H_2S exists in a non-dissociated, molecular state in the ionic liquid solvent, it is released as a gas from the reaction mixture (b.p. -82 °C), giving rise to a system in which a catalytic amount of iodine can be used to produce streams of hydrogen sulfide and sulfuric acid. From these products, hydrogen can be generated through steam reforming of the hydrogen sulfide at temperatures attainable by nuclear power or solar concentrators, regenerating sulfur dioxide, and the sulfuric acid stream can be processed as in the various sulfur based cycles. This novel cycle shown in the scheme below, which we term the Sulfur-Sulfur cycle, contains reactant streams that are strictly fluids, like the Sulfur-Iodine cycle, and may be amenable to simpler implementation than other current cycles.

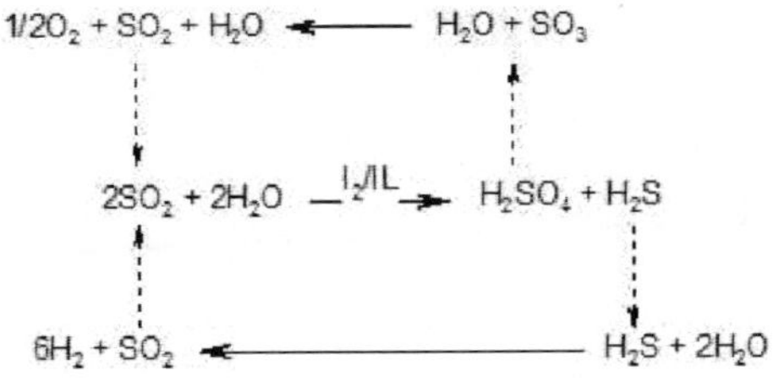

The presentation will focus on the new cycle concept, conditions favorable to the production of H_2S and the feasibility of steam reforming of the H_2S; in particular, results of hydrogen sulfide production in various ionic liquids under various conditions and reactant concentrations will be shown.

---

[1] Giaconia, A., *et al.* "Experimental study of two phase separation in the Bunsen section of the sulfur-iodine thermochemical cycle" *Intl. Jour. Hydrogen Energy* **2005**, *30*, 619-626.
[2] Vitart X, Carles P, Anzieu P. "A general survey of the potential and the main issues assoiciated with the sulfur iodine thermochemical cycle for hydrogen production using nuclear heat" *Prog Nuclear Energy* **2008**, *50*, 401-409.
[3] A. Yokochi, N. AuYeung, "Initial studies of the feasibility of performing the Bunsen reaction in a low vapor pressure liquid medium" *2008 Meeting of the AIChE*.
[4] Sakurai, M., *et al.* "Experimental study on side-reaction occurrence condition in the iodine-sulfur thermochemical hydrogen production process" *International Journal of Hydrogen Energy* **2000**, *25*, 613-619.
See more of this Session: Advances in Thermochemical Hydrogen Production

See more of this Group/Topical: Topical 8: Hydrogen Production for a Hydrogen Economy

Carbon Catalysts for Hydrogen Iodide (HI) Decompostion Inside the S-I Thermochemical Cycle

Paolo Favuzza, ENEA, Rome, Italy

Luigi Bimbi, ENEA, Rome, Italy

Claudio Felici, ENEA, Rome, Italy

Luigi Nardi, ENEA, Rome, Italy

Annarita Spadoni, ENEA, Rome, Italy

Pietro Tarquini, ENEA, Rome, Italy

Sulfur-iodine cycle is today one of the most widely investigated thermochemical cycles for hydrogen production. It consists of three reactions, whose sum is exactly water decomposition to give oxigen and hydrogen. Hydrogen iodide decomposition is one of these three reactions, the one, indeed, where hydrogen production takes place.

This gas phase reaction, rather insensitive from a thermodynamic point of view to variations of temperature, can anyway proceed faster if temperature is raised and using a proper catalyst. Various catalysts have been already considered for this application, particularly Nickel and Platinum, but also other transition metals, supporting them over alumina, ceria, carbon. Carbon alone has been recently investigated not only as support, but as a catalyst itself.

Our work is focused on the study of hydrogen iodide decomposition using commercial carbon catalysts in pellets (from NORIT) having different particle diameter and surface properties.

These catalysts gave us good results, providing conversion values, at 400-500°C, very close to thermodynamical expected ones, even at high spatial velocity. Also, they did not show any deactivation after more than several days of works. A kinetic mechanism was considered for them and the experimental results fitted on the basis of the related equations, allowing us to get the activation energy and the adsorption parameters.

Since the reaction was studied both feeding in the reactor the azeotropic hydriodic acid solution (57% by weigth of HI in water) - which is the usual solution coming from the other sections of S-I cycle - and feeding the pure hydrogen iodide gas, it was also possible to evaluate the effect of water on the efficiency of the reaction. Interference of little amounts of iodine or sulphur compounds (which could follow HI in the reactor as derived from the previous reactions of the cycle) was also considered.

The carbon catalysts investigated, due to the good results of our experimentation and to their low cost and availability, can therefore be considered as an interesting solution for HI decomposition, also for large-scale plant.

See more of this Session: Reaction and Separations for Thermochemical Cycles
See more of this Group/Topical: Topical 8: Hydrogen Production for a Hydrogen Economy

Catalyst Stability for Sulfuric Acid Decomposition as Influenced by Platinum Group Metal Composition and Reaction Pressure

Daniel M. Ginosar, Chemical Sciences, INL, Idaho Falls, ID

Lucia M. Petkovic, Chemical Sciences, INL, Idaho Falls, ID

Kyle C. Burch, Chemical Sciences, INL, Idaho Falls, ID

Thermochemical cycles produce hydrogen through a series of chemical reactions where the net result is the production of hydrogen and oxygen from water at much lower temperatures than from direct thermal decomposition. All chemicals within the cycle are recycled and the heat to drive the reactions, which tend to be endothermic, must be provided by a primary energy source. When the primary energy driver is nuclear or solar heat, hydrogen can be generated without producing greenhouse gasses, and can provide independence from our dwindling supplies of fossil fuels.

Among the high number of thermochemical water-splitting cycles proposed in the literature, the sulfur based cycles, including the Sulfur-Iodine (S-I) and the hybrid sulfur (HyS) cycles have generated considerable interest. The sulfur cycles all employ a high temperature catalytic sulfuric acid decomposition reaction to produce oxygen and generate SO2 for recycle. Platinum supported on porous titanium oxide has been shown to be very active over a wide range of temperatures; however, the catalytic metal deactivates too rapidly, primarily due to sintering in the oxidizing environment.

This paper will focus on the addition of high melting point platinum group metals to the platinum catalyst as an option to reduce deactivation. In addition, the affect of elevated pressure (up to 30 bar) on catalyst stability will be presented.

See more of this Session: Reaction and Separations for Thermochemical Cycles
See more of this Group/Topical: Topical 8: Hydrogen Production for a Hydrogen Economy

EXPERIMENTAL STUDIES OF THE BUNSEN REACTION IN THE SULFUR-IODINE PROCESS

Mariapaola Parisi, Alberto Giaconia#, Salvatore Sau#, Annarita Spadoni#, Giampaolo Caputo#, Pietro Tarquini#.*

** Chemical Engineering Department, University of Rome "La Sapienza", Rome, Italy*

 # ENEA, Rome, Italy

Abstract: this work deals with the investigation of the Bunsen reaction. An experimental campaign was conducted by means of a stirred reactor. Runs were carried out in a semibatch mode by feeding gaseous sulphur dioxide in a iodine/hydriodic acid/water solution. The purification of the produced hydriodic phase was also investigated. It was shown that both the amount of absorbed SO_2 and its conversion lowered as the temperature increased. Purification of the hydriodic phase was only possible at high temperature and high iodine contents.

Introduction

The Italian national agency for new technologies, energy and environment (ENEA), with the co-operation of the University of Rome "La Sapienza", is involved in a research investigation on thermochemical hydrogen production by means of the Sulfur-Iodine water splitting process (S-I). The aim of the research project is the production of hydrogen from solar energy with a higher yield and a reduced cost when compared with the traditional process consisting of water electrolysis connected to a photovoltaic device. This study reports the experimental work conducted on the reaction of sulphur dioxide with iodine and water to produce sulphuric and hydriodic acid in the S-I process:

$$2H_2O + SO_2 + I_2 \leftrightarrow H_2SO_4 + 2HI$$

According with the commonly adopted recipe [1], this reaction is carried out in liquid media under a certain excess of both water (n) and iodine (x) with respect to the stoichiometric values. By this means it is possible to thermodynamically drive the reaction and enhance separation of the two products into two corresponding liquid phases, namely a sulphuric acid rich phase and a "HIx phase". Therefore, almost all excess iodine is finally dissolved in the heavier HIx phase, while water will split between the two contacted liquid phases.

Accordingly, the process can be schematized as follows:

$$(2+n)H_2O + SO_2 + (1+x)I_2 \leftrightarrow [H_2SO_4 + (n-m)H_2O]_{Sulf.\ phase} + [2HI + mH_2O + xI_2]_{HIx\ phase}$$

The performance of the whole S-I process is very sensitive to the composition of the two phases. Moreover, we recently highlighted that a non-negligible amount of sulfur-containing compounds are finally dissolved in the HIx phase [2]. These sulphates (and residual SO_2) must be quantitatively removed in a downstream purification unit before subsequent operations for HI decomposition. This work deals with the experimental investigation of the Bunsen reaction and the purification of the HIx phase.

Experimental set up and procedures

The experimental setup, shown in Fig. 1, consists of a 500 mL jacketed, stirred reactor. The reactor is connected to an acid gas trap, filled with a small amount of sodium hydroxide solution, to absorb the unreacted sulphur dioxide. A sensor, placed on the trap exit line, detects the saturation of the alkaline solution. A thermostatic bath, connected to the reactor jacket, allows the regulation of the reactor temperature. The stirrer is connected to an electrical engine rotating at a speed equal to 500 rpm.

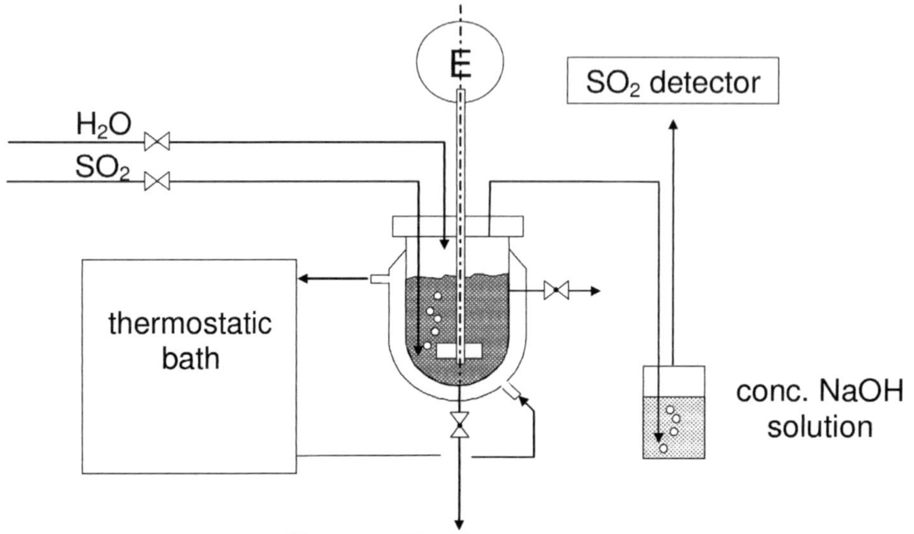

Figure 1. The Bunsen reactor.

The reaction was conducted by adopting a semi-batch procedure. The reactor was initially filled with iodine, hydriodic acid and water and the thermostatic bath and the stirrer were switched on. Once the desired temperature was attained, and iodine dissolved, gaseous SO_2 was fed at a rate of 83 nmL/min, until the SO_2 sensor detected the saturation of both the reactor and the alkaline trap. Then the stirrer was arrested and the mixture was kept motionless for 30 minutes to allow the stratification of the sulphuric and hydriodic phases: afterwards the two phases were separated, and the hydriodic one was sampled and analysed. The trap was weighted before and after the experiment in order to exactly determine the quantity of SO_2 absorbed in the reactor. For each run, the temperature was maintained constant at a fixed value within the range of 30-120°C.

A series of preliminary runs was conducted by adopting a different procedure: after the absorbed SO_2 reached the saturation concentration, additional water was fed for a short time. Then, the water addition was arrested and more SO_2 was added into the reactor. For this series of runs, the adopted temperatures were quite low, i.e. 28-35°C, and the composition of the phases was not analyzed, thus only qualitative results were obtained.

A fraction of the collected hydriodic phase, i.e. approximately 50 ml, was subjected to a stripping procedure in a thermostated bubble apparatus, shown in Fig. 2. Nitrogen was fed at a rate of 40 nmL/min for 30 minutes at constant temperature (the same previously adopted for the reaction). After the stripping, the residual sulphate content was measured. The stripping should remove all the sulphuric compounds, i.e. both the sulphur dioxide and the sulphuric acid, since Bunsen reaction is reversible:

$$H_2SO_4 + 2HI \rightarrow I_2 + 2H_2O + SO_2\uparrow$$

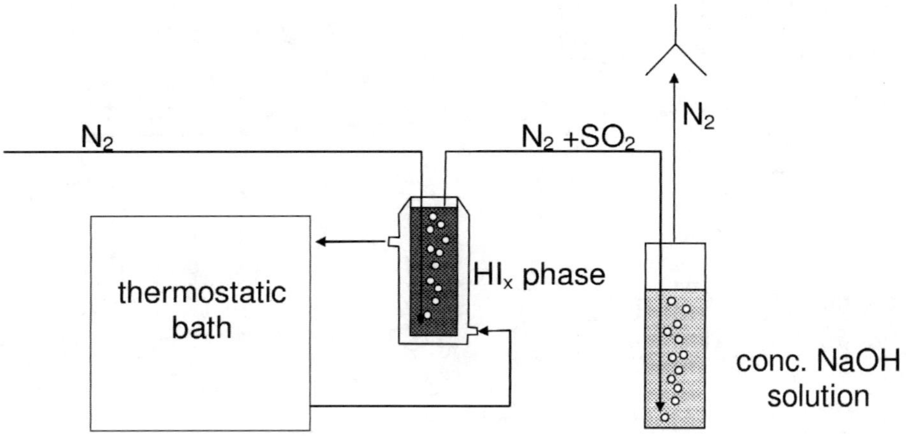

Figure 2. The stripping apparatus.

The samples from the HIx phase were first collected in an alkaline aqueous solution (NaOH 1 M), in order to let the iodine dismute. Sulfates, iodides and iodates contents were determined by ion chromatography: the original iodine content was determined by measuring the iodates, while the hydriodic acid concentration was calculated from the overall iodides content minus the amount derived from iodine dismutation.

Since the chromatographic analysis does not discriminate between SO_2 and H_2SO_4, the amount of reacted sulphur dioxide was inferred from the change in the concentrations of iodine and hydriodic acid before and after the run: considering the stoichiometry of the Bunsen reaction, for each mole of reacted SO_2 one mole of iodine is consumed and two moles of hydriodic acid are formed, leading to an increase in the ratio between the hydriodic acid and the iodine concentrations as the SO_2 reacts. The implicit equation that connects the molar fraction of I_2 and HI before and after the run, $x_{I2,0}$, $x_{HI,0}$, x_{I2} and x_{HI}, respectively, with the molar fraction of reacted SO_2, $x_{SO2,r}$, is:

$$\frac{x_{HI}}{x_{I2}} = \frac{x_{HI,0} + 2 \cdot x_{SO2,r}}{x_{I2,0} - x_{SO2,r}}$$

Results and discussion

The preliminary series of runs clearly showed that the stratification of the two phases is a slow process at low temperature and low iodine concentration. Moreover, the addition of water had two main effects. The first is that the separation of the phases became faster: the sulphuric phase always appeared upon water feeding. The second effect is the enhancement of the capacity of absorption for the SO_2: an increase of about 28% in the water content, allowed a comparable increase in the quantity of absorbed sulphur dioxide.

The adopted operating condition and the obtained results for the main experimental runs are reported in Tabs. 1 to 3: the last two columns of Tab. 3 refer to a couple of runs, conducted at 30 and 45°C, respectively, that were interrupted before the complete saturation was attained, in order to monitor the dynamic behaviour of the reaction.

ID / operating variables & results	A	B	C	D	E
T, °C	30	45	45	45	45
initial feed, mol	7.93	5.29	4.74	8.00	7.13
composition of the feed (molar fraction):					
H_2O	0.65	0.65	0.59	0.66	0.59
HI	0.12	0.12	0.11	0.12	0.11
I_2	0.22	0.22	0.30	0.22	0.29
absorbed SO_2, molar fraction	0.075	0.064	0.068	0.065	0.054
reacted SO_2, molar fraction	0.065	0.030	0.037	0.042	0.037
estimated conversion	0.87	0.47	0.55	0.66	0.68
composition after the reaction (molar fraction):					
H_2O	0.51	0.56	0.53	0.52	0.50
HI	0.26	0.19	0.18	0.23	0.19
I_2	0.16	0.21	0.25	0.20	0.26
sulphates	0.074	0.037	0.045	0.056	0.053
sulphates content after stripping		0.014		0.050	0.003

Table 1. Experimental runs from A to E (saturated runs).

ID / operating variables & results	F	G	H	I	J
T, °C	60	60	80	80	95
initial feed, mol	5.29	4.81	5.31	5.37	5.63
composition of the feed (molar fraction):					
H_2O	0.65	0.62	0.55	0.56	0.54
HI	0.12	0.13	0.11	0.11	0.11
I_2	0.22	0.25	0.34	0.34	0.36
absorbed SO_2, molar fraction	0.025	0.034	0.021	0.021	0.015
reacted SO_2, molar fraction		0.028	0.006		0.005
estimated conversion		0.82	0.30		0.33
composition after the reaction (molar fraction):					
H_2O		0.60	0.50	0.53	0.455
HI		0.17	0.13	0.11	0.133
I_2		0.20	0.36	0.35	0.402
sulphates		0.025	0.013	0.014	0.010
sulphates content after stripping			0.002	0.001	0.001

Table 2. Experimental runs from F to J (saturated runs).

ID / operating variables & results	K	L	M	*Y*	*Z*
T, °C	95	120	120	*30*	*45*
initial feed, mol	5.00	3.20	3.20	*5.29*	*7.8*
composition of the feed (molar fraction):					
H_2O	0.53	0.36	0.36	*0.65*	*0.65*
HI	0.11	0.07	0.07	*0.12*	*0.13*
I_2	0.36	0.57	0.57	*0.22*	*0.23*
absorbed SO_2, molar fraction	0.017	0.010	0.005	*0.036*	*0.025*
reacted SO_2, molar fraction	0.004	#	#	*0.035*	*0.020*
estimated conversion	0.26	#	#	*0.97*	*0.79*
composition after the reaction (molar fraction):					
H_2O	0.49	0.34	0.51	*0.62*	*0.60*
HI	0.12	0.07	0.09	*0.18*	*0.17*
I_2	0.38	0.59	0.40	*0.18*	*0.21*
sulphates	0.010	0.004	0.003	*0.026*	*0.020*
sulphates content after stripping	0.001	0.000	0.000	*0.016*	

Table 3. Experimental runs from K to M (saturated runs) and Y-Z (unsaturated runs).

In the F run (table 2), the hydriodic phase was not analysed, hence only the value for the absorbed SO_2 is reported. Concerning the L and M tests, conducted at 120°C, the formation of the sulphuric phase was observed, thus some sulphuric acid was surely formed. Nevertheless the very low amount of absorbed SO_2, coupled with the error on the compositions analysis, does not allow to obtain any result with physical sense concerning the amount of reacted SO_2 and, again, only the value for the absorbed SO_2 is reported.

It is important to point out that the iodine concentration and the temperature are not independent operating variables: the concentration of iodine at low temperature is limited by the solubility of iodine in hydriodic acid/water solutions; on the other hand, low concentration of iodine at high temperature could not be adopted, since it will lead to undesired secondary reactions [3].

In Fig. 3 the absorbed SO_2 in the Bunsen reactor is plotted Vs the operating temperature: the solid symbols refer to the runs where SO_2 saturation was attained and they show a clear trend with the temperature, despite the not negligible relative experimental error, i.e. about 10%. This effect can be partially due to the exothermicity of Bunsen reaction (equilibrium shifts towards SO_2 by increasing the temperature) and to the lower SO_2 solubility in the liquid.

A good fitting may be obtained with an exponential law (correlation coefficient higher than 94%).

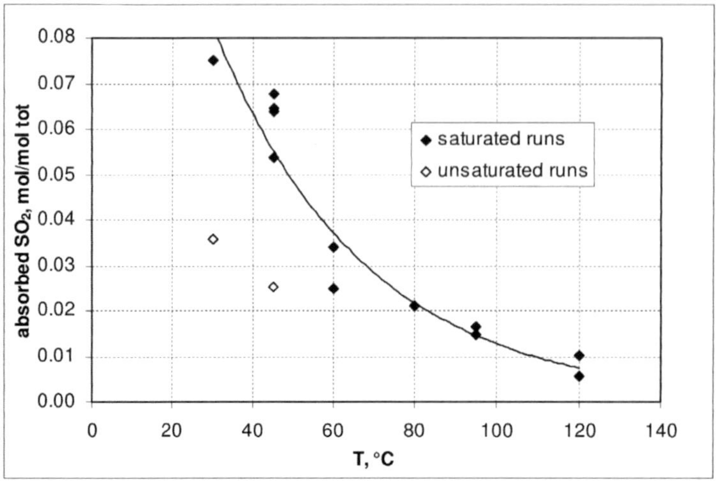

Figure 3. Absorbed SO_2 (molar fraction with respect to the final total moles) Vs temperature.

In Fig. 4 the reacted SO_2 is plotted Vs temperature. As previously stated, the amount of reacted SO_2 was inferred from the change in the ratio between the iodine and the hydriodic acid contents and the relative error resulting from this procedure is quite significant, i.e. approximately 15%: nevertheless the lowering effect of the temperature on the reacted SO_2 is clearly evident. A 3rd degree polynomial fitting curve led to a correlating coefficient higher than 97%.

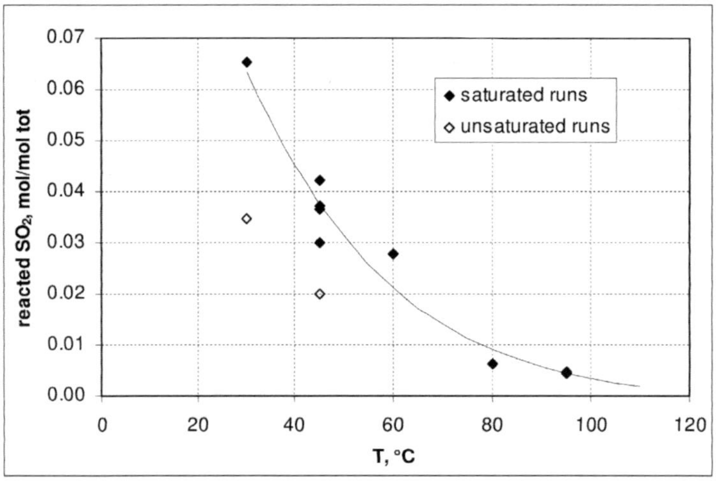

Figure 4. Estimated reacted SO_2 (molar fraction with respect to the final total moles) Vs temperature.

It is interesting to estimate what fraction of the absorbed sulphur dioxide actually reacts at different temperatures, that is, the effect of the temperature on the SO_2 conversion. A first attempt was made by calculating the ratio between the reacted and the absorbed SO_2 for each single run: unfortunately, this procedure leads to very scattered data, with an estimated relative error that can exceed 20%. In fact, both the reacted and the absorbed SO_2 measures are affected by not-negligible errors, that propagate when the ratio is computed, especially for very low values. A different approach consists in estimating the conversion as the ratio between the two fitting lines reported in Figs. 3 and 4, respectively: the obtained estimated trend is reported in Fig. 5 (solid line). It is interesting to notice that the estimated conversion of the run where SO_2 did not reached the saturation (open symbol in the diagram) is considerably higher.

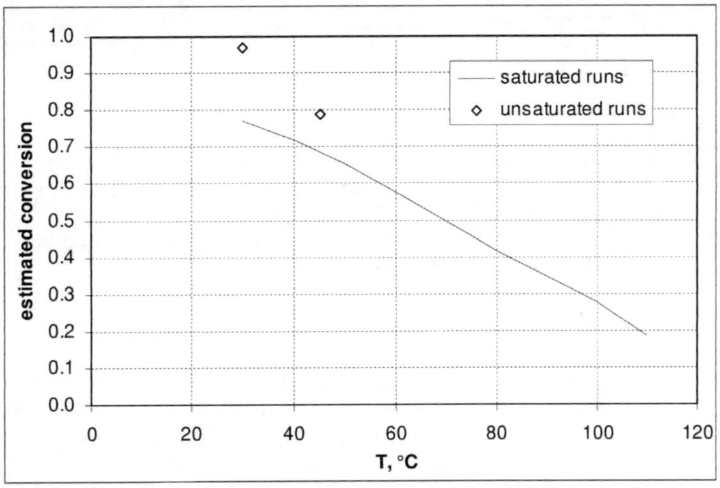

Figure 5. Estimated conversion Vs temperature.

Finally, the sulphates content in the HIx phase before and after the stripping is reported in Fig. 6. The measured sulphates contents before stripping are in good agreement with the ones previously measured in several investigation on the sulphuric/HIx phase LLE [2], [4], [5]. As clearly shown by the diagram, the stripping operation strongly lowered the sulphates content: nevertheless, sulphates are not quantitatively removed below 120°C. On the other hand, when the temperature is risen and the iodine molar fraction is below 0.35, secondary reactions take place, as reported by Sakurai et al. [3] and verified during the accomplishment of the experimental campaign here reported, as solid sulfur formation was observed in the stripping device.

Since the presence of sulphates in the HIx phases must be avoided in the following sections of the thermochemical cycle, it is necessary to adopt an high temperature for the Bunsen reactor, despise the low SO_2 solubility and the even lower conversion achievable.

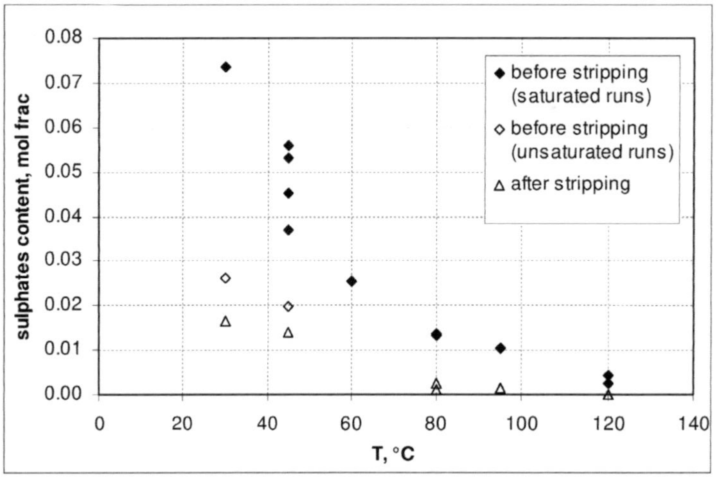

Figure 6. Sulphates content in the hydriodic phase.

Conclusion

The work here presented investigates the influence of the iodine concentration and the operating temperature on the Bunsen reaction: these two variables are not independent, since there are solubility limits for the iodine at low temperature and side reactions are presented if the iodine concentration is too low at high temperature.

At temperatures lower than 35°C, the separation of the two phases is quite slow, unless a great excess of water is adopted. Beside that, the effects of the temperature and the iodine content increases can be summarized as follows: both the solubility of the sulphur dioxide and its conversion decrease, the contamination of sulphates in the HIx phases decreases and the efficacy of the stripping purification step increases. Particularly, only when both the temperature and the iodine content are very high, it is possible to quantitatively remove the sulphates from the HIx phase without the occurrence of secondary reactions.

References

1. Brown LC, Besenbruch GE, Lentsch RD, Schultz KR, Funk JF, Pickard PS, et al. (2003), "High efficiency generation of hydrogen fuels using nuclear power", *General Atomic Project 30047.*
2. Giaconia A, Caputo G, Ceroli A, Diamanti M, Barbarossa V, Tarquini P, Sau S (2007), "Experimental study of two phase separation in the Bunsen section of the sulfur-iodine thermochemical cycle", *Int J Hydrogen Energy*, 32, pp. 531–536.
3. Sakurai M, Nakajima H, Amir R, Onuki K, Shimizu S (2000), "Experimental studies on side-reaction occurrence condition in the iodine–sulfur thermochemical hydrogen production process", *Int J Hydrogen Energy*, 25, pp. 613–619.
4. Giaconia A, Sau S, Caputo G, Tarquini P (2007), "Continuous Flow Operation of a Bunsen Reactorin the Iodine-Sulfur Thermochemical Water-Splitting Cycle", *AIChE 2007 Annual Meeting, Salt Lake City (UT).*
5. Byung Jin Lee, Hee Cheon NO, Ho Joon Yoon, Seung Jun Kim, Eung Soo Kim (2008), "An optimal operating window for the Bunsen process in the I–S thermochemical cycle", *Int J Hydrogen Energy*, 33, pp. 2000–2010.

Efficient Hydrogen Production with the Reformer Sponge Iron Cycle (RESC)

Markus Thaler, Institute of Chemical Engineering and Environmental Technology, Graz University of Technology, Graz, Austria

Viktor Hacker, Institute of Chemical Engineering and Environmental Technology, Graz University of Technology, Graz, Austria

Matthaeus Siebenhofer, Institute of Chemical Engineering and Environmental Technology, Graz University of Technology, Graz, Austria

The reformer sponge iron cycle (RESC) is a process, based on a thermochemical cycle operated at high temperatures. It is designed to convert hydrocarbons from renewable resources to hydrogen with a quality that exceeds the requirements of fuel cells. The RESC process consists of a conventional catalytic steam reformer unit where natural gas is split into a gas mixture containing H_2, CO, CO_2 and H_2O. In the subsequent sponge iron reaction (SIR) process the reducing components of the produced synthesis gas are converted into pure hydrogen. In a first step, CO and H_2 are used to reduce iron oxide to metallic iron (Eq.1+2). During the following oxidation with steam pure hydrogen is released (Eq. 3).

Reduction step:

$Fe_3O_4 + 4H_2 \rightarrow 3Fe + 4H_2O$ (Eq. 1)

$Fe_3O_4 + 4CO \rightarrow 3Fe + 4CO_2$ (Eq. 2)

Oxidation step:

$3Fe + 4H_2O \rightarrow Fe_3O_4 + 4H_2$ (Eq. 3)

The SIR process hence replaces the shift conversion as well as the final purification of the hydrogen rich gas. These steps conventionally consist of a combination of units for carbon dioxide removal, methanation, molecular sieves, purifiers, pressure swing adsorption (PSA) or liquid nitrogen scrubbing.

In order to accomplish a semi continuous work of the process more reactors have to be operated in parallel whereby the number of the reactors depends on the ratio of needed time for reduction to oxidation. Reformer unit and SIR operate at similar temperatures (approximately 800°C), which eases the system integration of both units. To achieve a higher efficiency 50% of the lean gas in the reduction step is recycled to the inlet of the reformer. Corresponding to the thermodynamic equilibrium in contact with the solid phase the lean gas contains CO_2, H_2O as well as non-reacted CO and H_2. With the recycle no or only a slight amount of additional steam is necessary for the reforming process.Depending on the fuel source up to 80% efficiency can be achieved.

The limitations of the SIR process are given by thermodynamic equilibria. Choosing iron as contact mass for the SIR has several advantages: iron is highly available, it has a relatively high energy density, it is cheap, non toxic and it can be recycled in the steel industry after consumption.

However, iron is deactivated rapidly after few reduction-oxidation cycles due to sintering of the metal species. The sintering causes a dramatic decrease of active surface area and the densification of the solid which leads to diffusion limitation of the gas-solid reactions. The addition of proper foreign metal oxides to the iron oxide enhances the cycle stability considerably.

The research activities of the Institute of Chemical Engineering and Environmental Technology concentrate on the improvement of the process efficiency and optimization of the contact mass. For this purpose studies on contact mass deactivation and lifetime improvement were conducted and a prototype test rig with two SIR reactors was installed.

See more of this Session: Reaction and Separations for Thermochemical Cycles
See more of this Group/Topical: Topical 8: Hydrogen Production for a Hydrogen Economy

H_2 GENERATION FROM WATER-SPLITTING USING NOVEL FERRITE FOAM MATERIALS

Rahul R. Bhosale, Rajesh V. Shende, Michael K. Opoku and Jan A. Puszynski,

Department of Chemical and Biological Engineering,
South Dakota School of Mines & Technology,
Rapid City, SD 57701 USA

Abstract

The ferrite materials containing metal cations such as Mn, Ni or Zn are considered to be potentially attractive for hydrogen generation using thermochemical water-splitting. In this paper $NiFe_2O_4$ was successfully synthesized using sol-gel technique. In this synthesis approach, Ni and Fe chlorides were used as precursors, ethanol as a solvent and propylene oxide (PO) as a proton scavenger. As-synthesized $NiFe_2O_4$ was further characterized using powder X-ray diffraction (XRD), BET surface area analyzer and scanning electron microscopy (SEM). Thermochemical water-splitting reactions were carried out using $NiFe_2O_4$ material in a packed bed reactor at 700°C and the material was regenerated at 900°C.

Introduction

As an ultimate clean fuel and its use in fuel cell, H_2 has tremendous potential as a substitute for fossil fuels [1]. The direct dissociation of water to produce H_2 is rather simple process, but in order to achieve significant conversion requires very high temperatures (>2200°C) [2]. Also, the product of this one step water splitting (thermolysis) is a mixture of H_2 and O_2. Generated H_2 and O_2 gases have to be separated at high temperatures, which imposes obstacle for industrial use because the separation process requires expensive membrane technology [3]. Therefore, this process has a very little chance of commercialization. On the other hand, thermochemical water splitting process, which utilize redox pairs of metal oxides have been regarded as the simplest, as well as most environmentally benign. This process produces hydrogen at lower temperature as compared to that of thermolysis and does not produce the explosive oxygen-hydrogen mixture.

The thermochemical water-splitting is a two-step process, which utilizes redox reaction of mixed valence metal oxides (MO) [4]. The two-step cycle proceeds with the endothermic reduction of a metal oxide at higher temperatures.

$$MO_{oxidized} \longrightarrow MO_{reduced} + \tfrac{1}{2} O_2 \qquad (1)$$

This reduced oxide reacts with water directly at a lower temperature and generates hydrogen, which resolves the $MO_{reduced}$ into $MO_{oxidized}$. The second step is slightly exothermic.

$$MO_{reduced} + H_2O \longrightarrow MO_{oxidized} + H_2 \qquad (2)$$

The two-step thermochemical process eliminates the formation of explosive H_2-O_2 mixture, avoiding the need for high temperature separation of oxygen and hydrogen. This process requires cycling of the water splitting and regeneration step. This cycling is disadvantageous because it is time consuming to heat and cool the redox material and complexities may arise due to the structural disintegration and potential sintering at higher temperatures.

The current research efforts on redox materials are focused on Fe_3O_4/FeO [5], Mn_3O_4/MnO [6], ZnO/Zn [7], and a combination of these or doped ferrites [3, 8]. Among these materials, ferrites such as $(A_xB_y)Fe_zO_4$ (doped spinel) or $A_xB_yFe_{1-x-y}O$ (doped wustite) or their mixtures with elements A and B being bivalent metal cations such as Mn, Ni or Zn [9] have been reported to be the most active materials for water splitting reaction [4]. The water splitting temperature was decreased up to 627°C for the Ni and Mn doped ferrites, but on the other hand the required temperature for regeneration was around 1300°C [10]. The difference between the two temperatures is around 673 °C which is too high.

Solid state synthesis, self-propagation high temperature synthesis method (SHS) and aerosol spray pyrolysis are some of the reported synthesis methods for ferrite materials [9]. Several ferrites synthesized by these methods are non-porous yielding low surface area. A low temperature sol-gel technique is a promising option available for ferrite preparation. Due to the requirement of low temperature for synthesis, the thermal degradation of the material is minimized. It is possible to prepare highly porous and nanocrystalline materials by this method. Due to the high porosity, the surface area will be more i.e. more number of active sites will be available for oxygen scavenging. In addition to this, sol-gel technique provides the control over the particle size, pore size, and the porosity of the final material.

In the present study, we report synthesis of Ni-ferrite material using sol-gel synthesis method and its characterization using powder X-ray diffraction, BET surface area analyzer and scanning electron microscopy (SEM). $NiFe_2O_4$ was used for hydrogen production at 700°C by water-splitting reaction and successfully regenerated at 900°C. By accomplishing this goal, a minimum temperature difference will be achieved. Also, an effort was given for the synthesis of foam-like material (having high surface area) using sol-gel synthesis approach involving the use of polymer microspheres with hydrocarbons present inside the core that are surrounded by polymeric shell.

Experimental Methods and Characterization

Ferrite Synthesis

The sol-gel method was employed for the synthesis of $NiFe_2O_4$ material. In the synthesis, Ni and Fe chlorides were dissolved in ethanol (solvent) by means of sonication. To this solution, propylene oxide (PO) was added as a gelation agent and the solution was left undisturbed for gel formation. The gel time depends upon the concentration of PO. To have foam-like morphology, the gel was mixed with 10 wt% polymer microspheres (Expancel, Azko Nobel). The polymer microsphere contains hydrocarbon gases encapsulated in a polymeric shell which is made up of thermoplastic material. As-synthesized gel was calcined at 900°C in N_2/air environment.

Ferrite Characterization

The sol-gel synthesized $NiFe_2O_4$ was characterized using powder X-ray diffraction, BET surface area analyzer and scanning electron microscopy (SEM).

Hydrogen Generation Set-up

The experimental set-up for thermochemical water splitting reaction and regeneration of $NiFe_2O_4$ is shown in Fig. 1. It consists of a high temperature inconel tubular reactor enclosed in a vertical split furnace (heating limit of 1200 °C, from Carbolite, England). The inconel reactor was packed with 10 g of $NiFe_2O_4$ with the help of rasching rings and quartz wool. A mass flow meter for N_2 (AALBORG, Inc. with flow range of 0-1000 smL min^{-1}) was placed for continuous monitoring of carrier gas flow. Distilled water (from the reservoir) was pumped with the help of peristaltic pump to the vaporizing and preheating furnace (heating limit of 1000 °C, from Carbolite, England) and the preheated steam was fed to the packed bed reactor. At the outlet of the reactor, a water adsorbent column was installed which contains drierite (anhydrous calcium sulfate, from W.A. Hammond Drierite company Inc., OHIO). A H_2 sensor was mounted on the outlet line to continuously monitor the exiting gas stream. Finally, the outlet gas stream was passed to flare.

Thermochemical water-splitting reaction was carried out at 700°C in both semi-batch and continuous modes. $NiFe_2O_4$ loaded packed bed reactor was heated up to 700°C under constant flow of N_2 (100 ml min^{-1}). Water was pumped to the vaporizing and preheating furnace (at 500°C), and the preheated steam was fed to the packed bed reactor. The effluent gas stream was continuously monitored using H_2 sensor. The sensor reading will indicate the H_2 (vol %) in the exit gas stream (which is a mixture of N_2 and H_2). The reacted material was regenerated at 900°C.

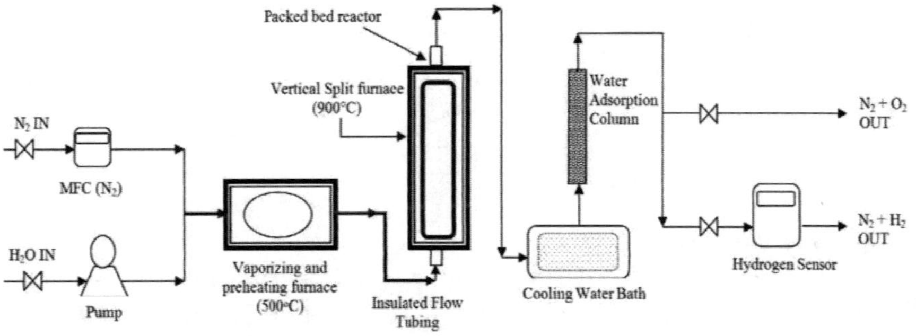

Fig.1 Experimental set-up for hydrogen generation.

Results and Discussion

Effect of PO concentration

In the sol-gel method, propylene oxide (PO) was added to the solution of Ni and Fe salts in ethanol. PO is a gelation agent and it was expected that the concentration of PO in the solution will affect the gel time. For the gel prepared with 2 g of precursor and 2 ml of PO, the gel time was 15 minutes. As the concentration of PO in the solution increases, the time required for gel formation decreases. The gel obtained was calcined at 900°C and the calcined material was analyzed using X-ray diffraction and BET surface area analyzer. The results obtained indicate that there is negligible effect on the SSA and the calcined phase composition, due to the increase in concentration of PO in the solution (Table1).

Table 1. Effect of PO concentration.

PO (ml)	SSA (m^2/g)	Phase
2	7.159	$NiFe_2O_4$
4	7.623	
6	7.558	
8	7.658	
10	7.644	

Effect of Aging Time

Aging time is an important parameter in the sol-gel synthesis because it may affect the pore size i.e. surface area. The synthesized gel was aged for 24 to 120 h and calcined at 900°C in air environment using muffle furnace. The calcined material was analyzed using BET surface area analyzer and powder X-ray diffraction. The results obtained are summarized in Table 2, which indicates no change in SSA and compositional purity.

Table 2. Effect of Aging Time.

Aging Time (h)	SSA (m^2/g)	Phase
24	7.259	$NiFe_2O_4$
48	7.158	
72	7.853	
96	7.982	
120	7.548	

Effect of Calcination Temperature

To study the effect of calcination temperature on the SSA and compositional phase, the sol-gel synthesized $NiFe_2O_4$ was calcined in the temperature range of 200 to 900°C in air environment using muffle furnace. The effect of calcination temperature on the SSA and crystalline phase formation is presented in Table 3. It was observed that with increasing calcination temperature the SSA decreases.

Table 3. Effect of Calcination Temperature.

Temperature (°C)	SSA (m^2/g)	Phase
200	84.698	$NiFe_2O_4$
400	74.485	
600	36.163	
800	12.989	
900	7.982	

X-ray diffraction pattern of the gel material calcined at different temperatures as shown in Fig. 2a, do not indicate the formation of other crystalline phases. To determine the particle size, the gel calcined at 900°C in air environment, was analyzed using scanning electron microscopy (SEM). The SEM image of the calcined $NiFe_2O_4$ is shown in Fig. 2b, showing smaller sized grains with the average particle size of < 1µm.

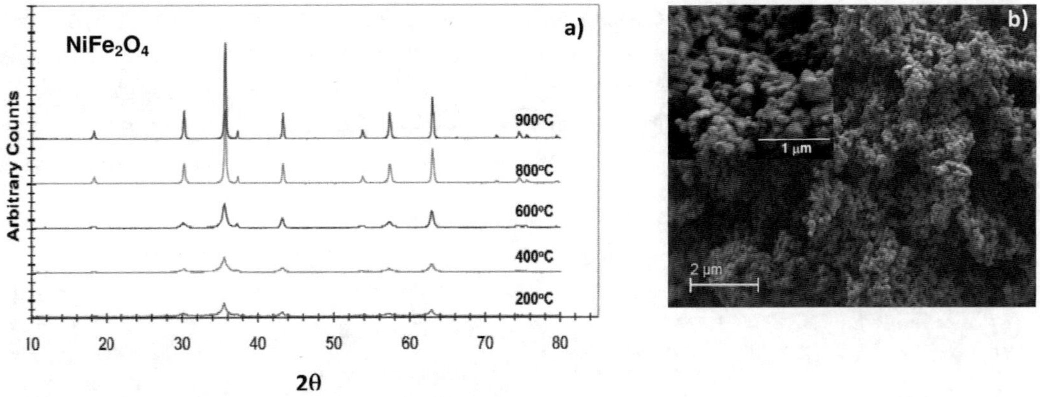

Fig. 2a) X-ray diffraction pattern of $NiFe_2O_4$ samples calcined at different temperatures and **b)** SEM image of $NiFe_2O_4$ calcined at 900°C in air environment.

Thermochemical Water Splitting

The water-splitting reactivity to produce hydrogen and regeneration ability of the sol-gel synthesized $NiFe_2O_4$ was studied in both semi-batch and continuous mode. In semi-batch process, a packed bed reactor loaded with 10 g of $NiFe_2O_4$ heated to 700°C in continuous flow of N_2 at 100 (ml min^{-1}). A batch of 10 ml of water was vaporized at 500°C using a vaporizing and preheating furnace and the preheated steam was fed into the packed bed reactor and H_2 was monitored continuously using a sensor. The hydrogen generation reaction was carried out in four batches without regenerating the $NiFe_2O_4$. The transient hydrogen profiles obtained for each batch are shown in Fig. 3. It was observed from the profiles that for each subsequent cycle the volume of hydrogen produced was less as compared to that of previous.

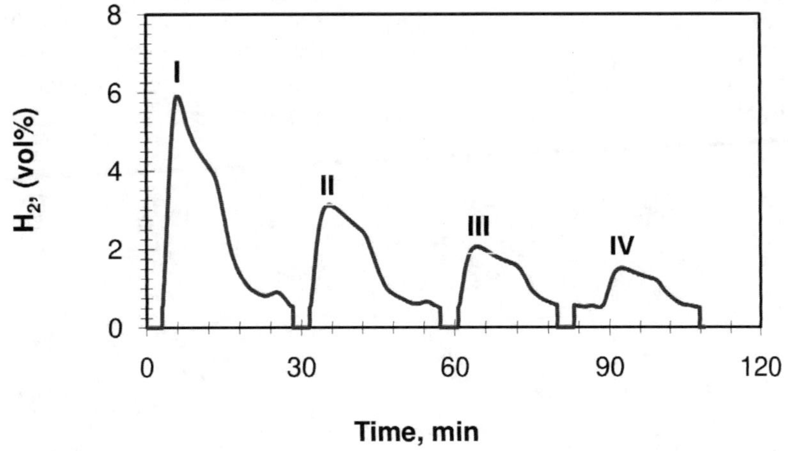

Fig.3 Transient hydrogen profiles for each batch in the semi-batch process.

In continuous process, the water was continuously pumped at the rate of 1ml min^{-1} from reservoir to the vaporizing and preheating furnace, which was held at 500°C. The preheated steam was fed to the packed bed reactor at 700°C and H_2 was monitored continuously using a sensor. After each water-splitting cycle, the material was regenerated at 900°C in N_2 environment for 1 h and reactor was cooled down again to 700°C to perform next cycle. Four such cycles (combination of water splitting and regeneration) were carried out. The transient hydrogen profile for the first and the second cycle is shown in Fig. 4a. This figure indicates that the duration of the first cycle was around 53 minutes which is 13 minutes more than that of the second cycle. Fig. 4b shows H_2 evolution during third and fourth cycles.

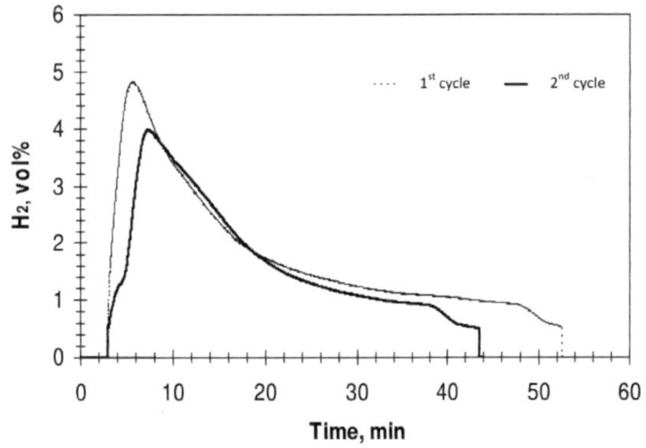

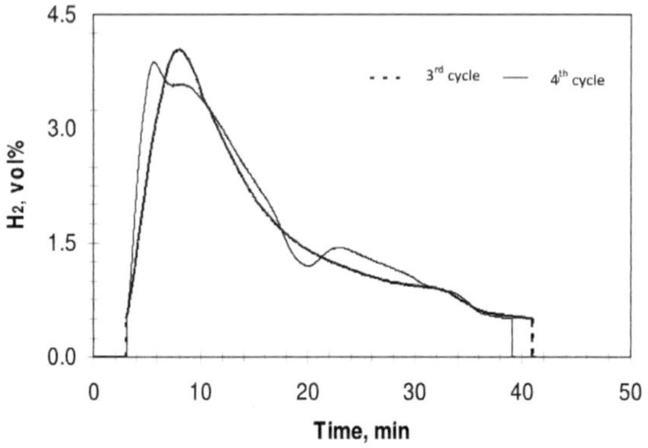

Fig.4 Transient hydrogen profile for continuous process **a)** 1st & 2nd cycles and **b)** 3rd & 4th cycles.

Effect of polymer microsphere addition on morphology

If ferrite is synthesized as a porous material, high surface area can be achieved. Synthesis method for ferrite with foam–like morphology was developed using the sol-gel technique where Ni & Fe salts were dissolved in ethanol by sonication and to this solution PO was added. After the gel formation, about 10 wt% polymer microspheres were added. A well-mixed gel containing these microspheres was fired rapidly in a tube furnace at about 900°C and quenched in N_2 environment. The SEM images of the calcined material are presented in Fig.5, which clearly exhibits foam-like morphology.

Fig.5 SEM images of calcined $NiFe_2O_4$ prepared using polymer microspheres.

The BET surface area of this foam-like material was greater than 70 m^2g^{-1} which is twice than that of Ni-Zn ferrites reported by others (36 m^2g^{-1}) [11].

Studies are in progress to understand how ferrite foam-like structures are formed during the rapid heating and quenching of the sol-gel synthesized material. We are working on the synthesis of desired crystalline phase i.e. $NiFe_2O_4$ in expancel added sample. Also, we are trying to understand the effectiveness of this foam-like material towards water-splitting reaction.

Conclusions

$NiFe_2O_4$ material was successfully synthesized using sol-gel technique. The effects of PO concentration, aging time and calcination temperatures on the SSA and crystalline phase were studied. Higher calcination temperature was found to reduce the SSA. Hydrogen generation using thermochemical water-splitting reaction with 10 g of $NiFe_2O_4$ packed in inconel tubular reactor was carried out in both semi-batch and continuous processes at 700°C. The material was successfully regenerated at 900°C and was reused at 700°C to perform water splitting reaction. A significant amount of H_2 was evolved in semi-batch as well as continuous processes. Sol-gel technique involving the addition of polymer microspheres in the ferrite gel produced foam-like material. The SSA of this foam-like material was about 70m^2g^{-1}, which is higher than that of the conventionally used ferrite materials in water-splitting reaction. Due to such a high surface area it is believed that the foam-like material will have tremendous potential for generating hydrogen from thermochemical water- splitting reaction.

Acknowledgement

The authors wish to acknowledge the financial support of National Science Foundation (NSF) for this work. (Grant No. CBET-0756214).

References

1. Farrell, A. E., Keith D.W., and Corbett, J.J. (2003), "A strategy for introducing hydrogen into transportation," *Energy Policy*, **31**, pp. 1357-1367.

2. Kodama, T. (2003), "High temperature solar chemistry for converting solar heat to chemical fuels," *Progress in Energy and Combustion Science*, **29** (6), pp. 567-597.

3. Tamaura, Y., Hasegawa, N., Kojima, M., Ueda, Y., Amano, H. and Tsuji, M. (1997), " Water splitting with the MN(II)-Ferrite-CAO-H_2O system at 1273K," *Energy*, **23**(10), pp. 879-886.

4. Shende, Rajesh V., Puszynski, Jan A., Opoku, Micheal K. and Bhosale, Rahul R. (2009), "Synthesis of novel-ferrite foam material for water-splitting application," *Nanotech ISBN 978-1-4398-1782-7*, **1**, pp. 201-204.

5. Steinfeld, A., Sanders, S. and Palumbo, R. (1999), "Design aspects of solar thermochemical engineering – A case study: Two-step water-splitting cycle using the Fe_3O_4/FeO redox system," *Solar Energy*, **65**(1), pp. 43-53.

6. Sturzenegger, M. and Nuesch, P. (1999), "Efficiency analysis for a manganese-oxide-based thermochemical cycle," *Energy*, **24** (11), pp. 959-970.

7. Steinfeld, A. (2002), "Solar hydrogen production via a two-step water-splitting thermochemical cycle based on Zn/ZnO redox reactions," *International Journal of Hydrogen Energy*, **27**(6), pp. 611-619.

8. Kojima, M., Sano, T., Wada, Y., Yamamoto, T., Tsuji, M. and Tamaura, Y. (1996), "Thermochemical decomposition of H_2O to H_2 on cation-excess ferrite", *Journal of Physics and Chemistry of Solids*, **57**(11), pp. 1757-1763.

9. Agrafiotis, C.C., Pagkoura, C., Lorentzou, S., Kostoglou, M. and Konstandopoulos, A.G. (2007), "Hydrogen production in solar reactors," *Catalysis Today*, **127**, pp. 265-277.

10. Tamaura, Y., Kojima, M., Ueda, Y., Hasegawa, N. and Tsuji, M. (1998) "Thermodynamic evaluation of water splitting by a cation-excessive (Ni, Mn) ferrite," *International Journal of Hydrogen Energy*, **23**(12), pp. 1185-1191.

11. Fu, Y.P. and Lin, C.H. (2002), "Microwave-Induced combustion synthesis of Ni-Zn ferrite Powder and its characterization," *Journal of Magnetism and Magnetic Materials*, **251**(1), pp. 74.

AUTHOR INDEX

American Institute of Chemical Engineers
3 Park Avenue
New York, New York 10016-5991

Systems Biology

Topical Conference at the 2008 AIChE Annual Meeting

Philadelphia, Pennsylvania
16 - 21 November 2008

ISBN: 978-1-61567-207-3

AIChE